Lecture Notes in Mathematics

Edited by A. Dold and B. Eckmann

566

Empirical Distributions and Processes

Selected Papers from a Meeting at
Oberwolfach, March 28 – April 3, 1976

Edited by P. Gaenssler and P. Révész

Springer-Verlag
Berlin · Heidelberg · New York 1976

Editors

Peter Gaenssler
Mathematisches Institut
Ruhr-Universität Bochum
Universitätsstraße 150
Gebäude NA
4630 Bochum/BRD

Pál Révész
Mathematical Institute of the
Hungarian Academy of Sciences
Réaltanoda Utca 13-15
1053 Budapest/Hungary

Library of Congress Cataloging in Publication Data

Main entry under title:

Empirical distributions and processes.

(Lecture notes in mathematics ; 566)
"Most of the papers in this volume were presented at the Oberwolfach-meeting on 'Mathematical stochastics'".
Bibliography: p.
Includes index.
1. Distribution (Probability theory)--Congresses. 2. Random variables--Congresses. 3. Convergence--Congresses. 4. Limit theorems (Probability theory)--Congresses. I. Gänssler, Peter. II. Révész, Pál. III. Series: Lecture notes in mathematics (Berlin) ; 566.
QA3.L28 no. 566 [QA273.6] 510'.8s [519.2'4]
76-30557

AMS Subject Classifications (1970): 60B10, 60F05, 60F15, 60G50, 62D05, 62E20, 62F10, 60F10, 60G15, 60G30, 60J65, 62E25

ISBN 3-540-08061-9 Springer-Verlag Berlin · Heidelberg · New York
ISBN 0-387-08061-9 Springer-Verlag New York · Heidelberg · Berlin

Printed in Germany

Printing and binding: Beltz Offsetdruck, Hemsbach/Bergstr.

PREFACE

Most of the papers in this volume were presented at the Oberwolfach-Meeting on "Mathematical Stochastics" held from March 28 to April 3, 1976.

This conference was planned with the intention of discussing recent developments in the theory of multivariate empirical processes, laws of iterated logarithm and invariance principles.

The lively discussions of all the participants on the interesting problems, the results and the various methods of proof, showed that this field of Mathematical Stochastics is of great interest and importance for both the probabilists as well as for the statisticians.

Especially for this volume we selected mainly those papers which were devoted to the investigation of empirical distributions and empirical processes.

We are especially grateful to Springer-Verlag for publishing the present contributions in the Lecture Notes Series.

P. Gaenssler, Bochum

P. Révész, Budapest

TABLE OF CONTENTS

Csörgő, M. and Burke, M.D.: Weak approximations of the empirical process when parameters are estimated 1

Csörgő, M. and Chan, A.H.C.: On the Erdös-Rényi increments and the P. Lévy modulus of continuity of a Kiefer process 17

Durbin, J.: Kolmogorov-Smirnov tests when parameters are estimated 33

Gaenssler, P. and Stute, W.: On uniform convergence of measures with applications to uniform convergence of empirical distributions 45

Krickeberg, K.: An alternative approach to Glivenko-Cantelli theorems 57

Neuhaus, G.: Weak convergence under contiguous alternatives of the empirical process when parameters are estimated: The D_k approach 68

Philipp, W.: Almost sure invariance principles for empirical distribution functions of weakly dependent random variables 83

Révész, P.: Three theorems of multivariate empirical process 106

Simons, G. and Stout, W.: Weak convergence to stable laws by means of a weak invariance principle 127

Stute, W.: A necessary condition for the convergence of the isotrope discrepancy 138

Topsøe, F., Dudley, R.M. and Hoffmann-Jørgensen, J.: Two examples concerning uniform convergence w.r.t. balls in Banach spaces 141

LIST OF PARTICIPANTS

O. Barndorff-Nielsen, Aarhus
H. Bauer, Erlangen
V. Baumann, Bochum
D. Bierlein, Regensburg
W.J. Bühler, Mainz
M. Csörgő, Ottawa
P.L. Davies, Münster
H. Dinges, Frankfurt
R.M. Dudley, Cambridge (USA)
D. Dugué, Paris
J. Durbin, London
F. Eicker, Dortmund
J. Fabius, Leiden
P. Gaenssler, Bochum
C. Grillenberger, Göttingen
F. Hampel, Zürich
H. Heyer, Tübingen
J. Hoffmann-Jørgensen, Aarhus
O. Kallenberg, Göteborg
H.G. Kellerer, München
D.G. Kendall, Cambridge
G. Kersting, Göttingen
J.F.C. Kingman, Oxford
K. Krickeberg, Paris
P. Major, Budapest
P. Mandl, Prag
D. Morgenstern, Hannover
U. Müller-Funk, Freiburg
M. Mürmann, Heidelberg
G. Neuhaus, Gießen
W. Philipp, Urbana
D. Plachky, Münster
P. Prinz, München
R.D. Reiss, Köln
P. Révész, Budapest
L. Rüschendorf, Aachen
K. Schürger, Heidelberg
V. Statulevicius, Vilnius
W. Stout, Urbana
J. Strobel, Bochum
W. Stute, Bochum
D. Szász, Budapest
J.D. Tacier, Zürich
H. von Weizsäcker, München
H. Witting, Freiburg
R. Zielinski, Warschau
W.R. van Zwet, Leiden

WEAK APPROXIMATIONS OF THE EMPIRICAL PROCESS WHEN PARAMETERS ARE ESTIMATED

M. Csörgő[1] and M.D. Burke[2]

Carleton University, Ottawa, Canada

0. Summary

Strong approximation results and methodology are used to obtain in-probability representations of the empirical process when the parameters of the underlying distribution function are estimated. These representations are obtained under a null hypothesis and a sequence of alternatives converging to the null hypothesis. The fairly general conditions on the estimators are often satisfied by maximum likelihood estimators. The asymptotic distribution of the estimated empirical process depends, in general, on the true value of the unknown parameters. Some useful methods of overcoming this difficulty are discussed.

1. Introduction.

Let $X_1, X_2, \ldots$ be a sequence of independent and identically distributed random variables (i.i.d.r.v.) with a continuous distribution function (d.f.) $F(x)$. Let $F_n(x)$ denote the proportion of $X_1, X_2, \ldots, X_n$ which are less than or equal to x $(x \in R)$, that is, let $F_n(x)$ be the empirical d.f.. The empirical process $\alpha_n(x)$ is defined by

$$\alpha_n(x) = \sqrt{n}\,[F_n(x) - F(x)]\ , \quad x \in R\ .$$

In [9], J. Komlós, P. Major and G. Tusnády have proved that $\alpha_n(x)$ is "near" to the following Gaussian processes:

(i) A Brownian Bridge $B(t)$ $(0 \le t \le 1)$, a Gaussian process with mean zero and covariance function $E\,B(s)\,B(t) = \min(s,t) - st$;

---

[1] Research partially supported by a Canadian N.R.C. Grant.

[2] Research supported by a Canadian N.R.C. Scholarship.

(ii) A Kiefer process $K(t,y)$ $(0 \leq t \leq 1 , 0 \leq y < \infty)$, a Gaussian process with mean zero and covariance function

$$E\, K(s,y_1)\, K(t,y_2) = \min(y_1,y_2)\, \{\min(s,t) - st\} .$$

For each fixed $y > 0$, $y^{-\frac{1}{2}} K(t,y) \overset{\mathcal{D}}{=} B(t)$.

We state their result.

Theorem A. (Komlós, Major, Tusnády [9]). If the underlying probability space is rich enough and if $F(\cdot)$ is a univariate continuous d.f., then one can define a sequence $\{B_n(\cdot)\}$ of Brownian Bridges and a Kiefer process $K(\cdot,\cdot)$ such that

$$\sup_{x \in R} |\, \alpha_n(x) - B_n(F(x))\, | \overset{a.s.}{=} \mathcal{O}\{n^{-\frac{1}{2}} \log n\}$$

and

$$\sup_{x \in R} |\, \sqrt{n}\, \alpha_n(x) - K(F(x),n)\, | \overset{a.s.}{=} \mathcal{O}\{\log^2 n\} .$$

Remark 1. By the phrase "if the underlying probability space is rich enough", we mean that an independent sequence of Wiener processes, independent of the i.i.d. sequence $\{X_n\}$, can be constructed on the assumed probability space. In the sequel, it will be assumed that the underlying probability spaces are rich enough in this sense.

Notations. By $\sup_x |\, h_n(x)\, | \overset{a.s.}{=} \mathcal{O}\{g(n)\}$, we mean that $\limsup_{n\to\infty} \sup_x |h_n(x)\, g(n)^{-1}| < \infty$ almost surely. The transpose of a vector $\underline{v}$ will be denoted by $\underline{v}^t$.

$\int \underline{v}$ will denote the vector $(\int v_1, \int v_2, \ldots, \int v_p)^t$, where $\underline{v} = (v_1, v_2, \ldots, v_p)^t$.

For a scalar function $f(\underline{\theta})$ of the vector $\underline{\theta}$, $(\partial/\partial\, \underline{\theta})\, f(\underline{\theta}_o)$ will denote the vector of partial derivatives of f with respect to the components of $\underline{\theta}$, evaluated at $\underline{\theta} = \underline{\theta}_o$.

Theorem A is useful for goodness-of-fit statistical tests when $F(\cdot)$ is completely specified. However, in most cases only the form of F is known while some parameters of F are not specified and must be estimated.

Consider the family of d.f. given by $\{H(x;\underline{\beta}_\circ,\underline{\theta}): \underline{\theta}\in\Omega\}$, where $\underline{\beta}_\circ$ is a p_1-dimensional column vector of specified parameters and $\underline{\theta}$ is a p_2-dimensional column vector of unknown parameters, but known to belong to a subset Ω of R^{p_2}. One is often interested in testing whether the true d.f. F of the i.i.d. sequence $\{X_n\}$ belongs to $\{H(x;\underline{\beta}_\circ,\underline{\theta}): \underline{\theta}\in\Omega\}$, that is, we wish to test the following null hypothesis:

(1) $$H_\circ : F \in \{H(x;\underline{\beta}_\circ,\underline{\theta}): \underline{\theta}\in\Omega\} .$$

Suppose that the vector $\underline{\theta}$ is estimated by $\{\hat{\underline{\theta}}_n\}$, a sequence of estimators, where each $\hat{\underline{\theta}}_n$ is obtained from the sample $X_1, X_2, \ldots, X_n$. Let

(2) $$\hat{\alpha}_n(x) = \sqrt{n}[F_n(x) - H(x;\underline{\beta}_\circ,\hat{\underline{\theta}}_n)]$$

be the estimated empirical process. We shall obtain "in probability" representations of $\hat{\alpha}_n(x)$ under the null hypothesis $H_\circ$ defined by (1) and also under a sequence of alternative hypotheses.

Under $H_\circ$, we shall assume that

(C1) $$\sqrt{n}\,(\hat{\underline{\theta}}_n - \underline{\theta}_\circ) = n^{-\frac{1}{2}} \sum_{j=1}^{n} \ell(X_j,\underline{\beta}_\circ,\underline{\theta}_\circ) + \epsilon_{\circ n} ,$$

where $\underline{\theta}_\circ$ is the true unknown value of $\underline{\theta}$ under $H_\circ$ and that the following conditions are satisfied: for a random observation x,

(i) $E\{\ell(x,\underline{\beta}_\circ,\underline{\theta}_\circ) \mid H_\circ\} = \underline{0}$;

(ii) $\epsilon_{\circ n} \xrightarrow{p} \underline{0}$ as $n\to\infty$;

(iii) $L(\underline{\beta}_\circ,\underline{\theta}_\circ) = E\{\ell(x,\underline{\beta}_\circ,\underline{\theta}_\circ)\,\ell(x,\underline{\beta}_\circ,\underline{\theta}_\circ)^t \mid H_\circ\}$ is a non-negative definite matrix;

(iv) ℓ is a continuous function of $(\underline{\beta},\underline{\theta})^t \in \nu$, uniformly in x;

(v) The derivative $(d/dx)\,\ell(x,\underline{\beta},\underline{\theta})$ is bounded uniformly in $x\in R$ and $(\underline{\beta},\underline{\theta})^t \in \nu$, where ν is the closure of a given neighbourhood of $(\underline{\beta}_\circ,\underline{\theta}_\circ)^t$.

We wish to study $\hat{\alpha}_n(x)$ under the sequence of alternatives given by

(3) $$A_n : F \in \{G_n(x;\underline{\beta}_n,\underline{\theta}): \underline{\theta}\in\Omega\} ,$$

where the family $\{G_n(x;\underline{\beta}_n,\underline{\theta}): \underline{\theta} \in \Omega\}$ has the same parametric structure as that of H of (1) and $G_n(x;\underline{\beta}_\circ,\underline{\theta}_\circ)$ converges uniformly to $H(x;\underline{\beta}_\circ,\underline{\theta}_\circ)$. For example, $\underline{\beta}$ and/or $\underline{\theta}$ could be location or scale parameters of the d.f. H and G_n.

Concerning the sequence of alternatives $\{A_n\}$, we will assume that

(C2) (i) $\sqrt{n}\,(\underline{\beta}_n - \underline{\beta}_\circ) \to \gamma$ as $n \to \infty$, for some p_1-dimensional vector γ,

(ii) $\sqrt{n}\,(\underline{\theta}_n - \underline{\theta}_\circ)$ is bounded uniformly in n, where $\underline{\theta}_n$ is the true unknown value of $\underline{\theta}$ under A_n, and that under the sequence of alternatives $\{A_n\}$:

$$\sqrt{n}\,(\hat{\underline{\theta}}_n - \underline{\theta}_n) = n^{-\frac{1}{2}} \sum_{j=1}^{n} \ell_n(X_j,\underline{\beta}_n,\underline{\theta}_n) + A\cdot\gamma + \epsilon_{1n},$$

where the p_2-vector functions ℓ_n are measurable and bounded (in n, x, and $(\underline{\beta},\underline{\theta})^t \in \nu$) and for a random observation x,

(iii) $E\{\ell_n(x,\underline{\beta}_n,\underline{\theta}_n)|A_n\} = \underline{0}$;

(iv) A is a given finite matrix of order $p_2 \times p_1$;

(v) $\epsilon_{1n} \overset{p}{\to} \underline{0}$, as $n \to \infty$;

(vi) the matrix $L_n(\underline{\beta}_n,\underline{\theta}_n) = E\{\ell_n(x,\underline{\beta}_n,\underline{\theta}_n)\,\ell_n(x,\underline{\beta}_n,\underline{\theta}_n)^t|A_n\}$ is non-negative definite and converges to $L(\underline{\beta}_\circ,\underline{\theta}_\circ)$ as $n \to \infty$;

(vii) ℓ_n is a continuous function of $(\underline{\beta},\underline{\theta})^t \in \nu$, uniformly in x;

(viii) the derivative $(d/dx)\,\ell_n(x,\underline{\beta},\underline{\theta})$ is bounded uniformly in $x \in R$, $(\underline{\beta},\underline{\theta})^t \in \nu$, and n;

(ix) $\sup_x ||\ell_n(x,\underline{\beta}_n,\underline{\theta}_n) - \ell(x,\underline{\beta}_\circ,\underline{\theta}_\circ)|| \to 0$ as $n \to \infty$, where $||\ \ ||$ is the usual Euclidean norm.

Conditions (C2) (i) to (vi), with $G_n = H$, $\ell_n = \ell$ and $\underline{\theta}_n = \underline{\theta}_\circ$, were the (A1) assumptions studied by Durbin [7]. An example which motivates our approach is

Example 1. Maximum likelihood estimators often satisfy conditions (C1) and (C2). Under fairly general regularity conditions, a sequence $\{\hat{\underline{\theta}}_n\}$ of maximum likelihood estimators satisfies

(4) $$\sqrt{n}\,(\hat{\underline{\theta}}_n - \underline{\theta}_\circ) = n^{-\frac{1}{2}} \sum_{j=1}^{n} (\partial/\partial\, \underline{\theta}) \log h(X_j;\underline{\beta}_\circ,\underline{\theta}_\circ)^t \cdot I^{-1}(\underline{\theta}_\circ) + \epsilon_{\circ n}$$

under $H_\circ$, where $I(\underline{\theta}_\circ)$ is the Fisher information matrix evaluated at $\underline{\theta} = \underline{\theta}_\circ$ and h is the density of H. Under A_n, the sum representation would be (cf. Durbin [7])

(5) $$\sqrt{n}\,(\hat{\underline{\theta}}_n - \underline{\theta}_n) = n^{-\frac{1}{2}} \sum_{j=1}^{n} (\partial/\partial\, \underline{\theta}) \log g_n(x;\underline{\beta}_n,\underline{\theta}_n)^t \cdot I_n^{-1}(\underline{\theta}_n)$$

$$+ I^{-1}(\underline{\theta}_\circ)\cdot I_{21}\cdot \gamma + \epsilon_{1n},$$

where $I_{21} = E\{(\partial/\partial\,\underline{\theta}) \log h(X_1;\underline{\beta}_\circ,\underline{\theta}_\circ) \cdot (\partial/\partial\,\underline{\beta}) \log h(X_1;\underline{\beta}_\circ,\underline{\theta}_\circ)|H_\circ\}$, g_n is the density of G_n and $I_n(\underline{\theta})$ is the Fisher information matrix with respect to $g_n(x;\underline{\beta}_n,\underline{\theta})$.

Concerning the distribution functions and the nature of the alternatives involved, we will assume the following:

(C3) (i) the d.f. $H(x;\underline{\beta},\underline{\theta})$ and $G_n(x;\underline{\beta},\underline{\theta})$ are uniformly continuous in $x \in R$ and $(\underline{\beta},\underline{\theta})^t \in \nu$;

(ii) the vectors of partial derivatives

$$(\partial/\partial(\underline{\beta},\underline{\theta})^t)\, G_n(x;\underline{\beta},\underline{\theta}) \quad \text{and} \quad (\partial/\partial(\underline{\beta},\underline{\theta})^t)\, H(x;\underline{\beta},\underline{\theta})$$

exist, are continuous functions of $(\underline{\beta},\underline{\theta})^t \in \nu$ uniformly in $x \in R$, and are uniformly bounded in $x \in R$, $(\underline{\beta},\underline{\theta})^t \in \nu$ and n;

(iii) for $\underline{\theta}$ satisfying $(\underline{\beta}_\circ,\underline{\theta})^t \in \nu$,

$$\sup_x |\sqrt{n}\,[G_n(x;\underline{\beta}_\circ,\underline{\theta}) - H(x;\underline{\beta}_\circ,\underline{\theta})] - w(x;\underline{\theta})| \to 0$$

as $n \to \infty$, where $w(x;\underline{\theta})$ is bounded and a continuous function of $\underline{\theta}$, $(\underline{\beta}_\circ,\underline{\theta})^t \in \nu$, uniformly in $x \in R$;

(iv) $$\sup_x \|\, (\partial/\partial(\underline{\beta},\underline{\theta})^t)\, G_n(x;\underline{\beta}_\circ,\underline{\theta}_\circ) - (\partial/\partial(\underline{\beta},\underline{\theta})^t)\, H(x;\underline{\beta}_\circ,\underline{\theta}_\circ))\,\| \to 0$$

as $n \to \infty$.

2. Weak Approximations of the Estimated Empirical Process.

Our main results are as follows:

Theorem 1. Under the sequence $\{A_n\}$ of alternatives, defined by (3), and assuming that conditions (C2) and (C3) are satisfied, one can construct a Gaussian process $\mathcal{G}(x,n;G_n,\underline{\beta}_n,\underline{\theta}_n)$ such that

$$\sup_{x\in R} | \hat{\alpha}_n(x) - \mathcal{G}(x,n;G_n,\underline{\beta}_n,\underline{\theta}_n) | \xrightarrow{P} 0 \quad \text{as} \quad n\to\infty, \tag{6}$$

where $\mathcal{G}$ is defined by

$$\begin{aligned}\mathcal{G}(x,n;G_n,\underline{\beta}_n,\underline{\theta}_n) &= n^{-\frac{1}{2}}K(G_n(x;\underline{\beta}_n,\underline{\theta}_n),n) \\ &- \left\{\int \ell_n(x,\underline{\beta}_n,\underline{\theta}_n)\, d_x\, n^{-\frac{1}{2}}K(G_n(x;\underline{\beta}_n,\underline{\theta}_n),n)\right\}^t \cdot \\ &\quad (\partial/\partial\, \underline{\theta})\, G_n(x;\underline{\beta}_n,\underline{\theta}_n) \\ &- (A\cdot\gamma)^t\cdot(\partial/\partial\, \underline{\theta})\, G_n(x;\underline{\beta}_n,\underline{\theta}_n) \\ &+ \gamma^t\cdot(\partial/\partial\, \underline{\beta})\, G_n(x;\underline{\beta}_n,\underline{\theta}_n) + w(x;\underline{\theta}_n),\end{aligned} \tag{7}$$

and $K(\cdot,\cdot)$ is the Kiefer process of Theorem A.

Corollary 1. Under $\{A_n\}$ and assuming that (C2) and (C3) are satisfied,

$$\sup_{x} | \hat{\alpha}_n(x) - \mathcal{G}(x,n;H,\underline{\beta}_\circ,\underline{\theta}_\circ) | \xrightarrow{P} 0 \quad \text{as} \quad n\to\infty,$$

where $\mathcal{G}(x,n;H,\underline{\beta}_\circ,\underline{\theta}_\circ)$ is defined by

$$\begin{aligned}\mathcal{G}(x,n;H,\underline{\beta}_\circ,\underline{\theta}_\circ) &= n^{-\frac{1}{2}}K(H(x;\underline{\beta}_\circ,\underline{\theta}_\circ),n) \\ &- \left\{\int \ell(x,\underline{\beta}_\circ,\underline{\theta}_\circ)\, d_x\, n^{-\frac{1}{2}}K(H(x;\underline{\beta}_\circ,\underline{\theta}_\circ),n)\right\}^t \cdot \\ &\quad (\partial/\partial\, \underline{\theta})\, H(x;\underline{\beta}_\circ,\underline{\theta}_\circ) \\ &- (A\cdot\gamma)^t\cdot(\partial/\partial\, \underline{\theta})\, H(x;\underline{\beta}_\circ,\underline{\theta}_\circ) \\ &+ \gamma^t\cdot(\partial/\partial\, \underline{\beta})\, H(x;\underline{\beta}_\circ,\underline{\theta}_\circ) + w(x;\underline{\theta}_\circ),\end{aligned} \tag{8}$$

and $K(\cdot,\cdot)$ is the Kiefer process of Theorem A.

Remark 2. In the case where $G_n = H$, $\ell_n = \ell$ and $\underline{\theta}_n = \underline{\theta}_0$, that is, under the Durbin sequence of alternative hypothesis, the resultant Gaussian process is defined by (8) but with $w(x;\underline{\theta}_0)$ equal to zero. Note also that although the true unknown value $\underline{\theta}_n$ of $\underline{\theta}$ under A_n is different from the true unknown value $\underline{\theta}_0$ under H_0, it does not affect the resultant Gaussian process, if (C2) (ii) is satisfied.

Theorem 2. Under the null hypothesis H_0, defined by (1), and assuming that (C1) and (C3) (i) and (ii) (concerning H) are satisfied,

$$\sup_x |\hat{\alpha}_n(x) - Q_0(x,n)| \overset{P}{\to} 0 \quad \text{as} \quad n\to\infty,$$

where $Q_0(x,n)$ is defined by (8) but with γ and $w(x;\underline{\theta}_0)$ equal to zero.

We shall first prove

Lemma 1. For the Kiefer process $K(\cdot,\cdot)$ of Theorem A, let

$$\epsilon_{2n}(x) = \sqrt{n}\,[F_n(x) - G_n(x;\underline{\beta}_n,\underline{\theta}_n)] - n^{-\frac{1}{2}}K(G_n(x;\underline{\beta}_n,\underline{\theta}_n),\ n).$$

Then under the sequence $\{A_n\}$ of alternatives,

$$\int \ell_n(x,\underline{\beta}_n,\underline{\theta}_n)\, d_x\, \epsilon_{2n}(x) \overset{a.s.}{\to} \underline{0} \quad \text{as} \quad n\to\infty.$$

Proof of Lemma 1. Evaluating the vector of stochastic integrals

$$\int \ell_n(x,\underline{\beta}_n,\underline{\theta}_n)\, d_x\, \epsilon_{2n}(x) = \lim_{N\to\infty} \{[\ell_n(x,\underline{\beta}_n,\underline{\theta}_n)\cdot \epsilon_{2n}(x)]_{x=-N}^{N}$$

$$- \int_{-N}^{N} (d/dx)\, \ell_n(x,\underline{\beta}_n,\underline{\theta}_n)\cdot \epsilon_{2n}(x)\, dx\}.$$

By (C2), ℓ_n and its derivative are uniformly bounded in $(\underline{\beta}_n,\underline{\theta}_n)^t \in \nu$, $x\in R$ and n. Since $\sup_x |\epsilon_{2n}(x)| \overset{a.s.}{=} O\{n^{-\frac{1}{2}}\log^2 n\}$, on letting $N = \log n$, the result follows.

<u>Proof of Theorem 1.</u> Under the sequence $\{A_n\}$ of alternatives, using Theorem A and the Taylor expansion of $G_n(x;\underline{\beta},\underline{\theta})$ and $H(x;\underline{\beta},\underline{\theta})$, we obtain

$$\hat{\alpha}_n(x) = \sqrt{n}\,[F_n(x) - H(x;\underline{\beta}_\circ,\hat{\underline{\theta}}_n)]$$

$$= \sqrt{n}\,[F_n(x) - G_n(x,\underline{\beta}_n,\underline{\theta}_n)] + \sqrt{n}\,[G_n(x;\underline{\beta}_n,\underline{\theta}_n)$$

$$- G_n(x;\underline{\beta}_\circ,\underline{\theta}_\circ)] + \sqrt{n}\,[G_n(x;\underline{\beta}_\circ,\underline{\theta}_\circ) - H(x;\underline{\beta}_\circ,\underline{\theta}_\circ)]$$

$$- \sqrt{n}\,[H(x;\underline{\beta}_\circ,\underline{\theta}_n) - H(x;\underline{\beta}_\circ,\underline{\theta}_\circ)]$$

$$- \sqrt{n}\,[H(x,\underline{\beta}_\circ,\hat{\underline{\theta}}_n) - H(x;\underline{\beta}_\circ,\underline{\theta}_n)]$$

$$= n^{-\frac{1}{2}}K(G_n(x;\underline{\beta}_n,\underline{\theta}_n), n) + \epsilon_{3n}(x)$$

$$+ \sqrt{n}\,(\underline{\beta}_n - \underline{\beta}_\circ)^t\cdot(\partial/\partial\,\underline{\beta})\,G_n(x;\underline{\beta}'_n,\underline{\theta}'_n)$$

$$+ \sqrt{n}\,(\underline{\theta}_n - \underline{\theta}_\circ)^t\cdot(\partial/\partial\,\underline{\theta})\,G_n(x;\underline{\beta}'_n,\underline{\theta}'_n)$$

$$+ w(x;\underline{\theta}_\circ) - \sqrt{n}\,(\underline{\theta}_n - \underline{\theta}_\circ)^t\cdot(\partial/\partial\,\underline{\theta})\,H(x;\underline{\beta}_\circ,\underline{\theta}''_n)$$

$$- \sqrt{n}\,(\hat{\underline{\theta}}_n - \theta_n)^t\cdot(\partial/\partial\,\underline{\theta})\,H(x;\underline{\beta}_\circ,\underline{\theta}^*_n)$$

$$= n^{-\frac{1}{2}}K(G_n(x;\underline{\beta}_n,\underline{\theta}_n), n) + \gamma^t\cdot(\partial/\partial\,\underline{\beta})\,G_n(x;\underline{\beta}'_n,\underline{\theta}'_n)$$

$$+ w(x;\underline{\theta}_\circ) - \{n^{-\frac{1}{2}}\sum_{j=1}^{n}\ell_n(X_j;\underline{\beta}_n,\underline{\theta}_n) + A\cdot\gamma\}^t\cdot(\partial/\partial\,\underline{\theta})\,H(x;\underline{\beta}_\circ,\underline{\theta}^*_n)$$

$$+ \epsilon_{4n}(x),$$

where $\epsilon_{3n}(x) = \epsilon_{2n}(x) + \sqrt{n}\,[G_n(x;\underline{\beta}_\circ,\underline{\theta}_\circ) - H(x;\underline{\beta}_\circ,\underline{\theta}_\circ)] - w(x;\underline{\theta}_\circ)$,

$$\epsilon_{4n}(x) = \epsilon_{3n}(x) + [\sqrt{n}\,(\underline{\beta}_n - \underline{\beta}_\circ) - \gamma]^t\cdot(\partial/\partial\,\underline{\beta})\,G_n(x;\underline{\beta}'_n,\underline{\theta}'_n)$$

$$+ \sqrt{n}\,(\underline{\theta}_n - \underline{\theta}_\circ)^t\cdot[(\partial/\partial\,\underline{\theta})\,G_n(x;\underline{\beta}'_n,\underline{\theta}'_n) - (\partial/\partial\,\underline{\theta})\,H(x;\underline{\beta}_\circ,\underline{\theta}''_n)]$$

$$+ \epsilon^t_{1n}\cdot(\partial/\partial\,\underline{\theta})\,H(x;\underline{\beta}_\circ,\underline{\theta}^*_n), \text{ and}$$

(9) $$||\underline{\beta}_n' - \underline{\beta}_\circ|| \leq ||\underline{\beta}_n - \underline{\beta}_\circ|| \text{ , } ||\underline{\theta}_n' - \underline{\theta}_\circ|| \leq ||\underline{\theta}_n - \underline{\theta}_\circ|| \text{ ,}$$

$$||\underline{\theta}_n'' - \underline{\theta}_\circ|| \leq ||\underline{\theta}_n - \underline{\theta}_\circ|| \text{ and } ||\underline{\theta}_n^* - \underline{\theta}_n|| \leq ||\hat{\underline{\theta}}_n - \underline{\theta}_n||$$

($||\cdot||$ is the usual Euclidean norm on R^p).

By (C3) (iii) and Theorem A, $\sup_x |\epsilon_{3n}(x)| \overset{p}{\to} 0$. By (C2) (i) and the continuity and boundedness properties (C3) (ii) of the partial derivatives,

$$\sup_x |[\sqrt{n}(\underline{\beta}_n - \underline{\beta}_\circ) - \gamma]^t \cdot (\partial/\partial \underline{\beta})\, G_n(x;\underline{\beta}_n',\underline{\theta}_n')| \to 0 \text{ . as } n \to \infty \text{ .}$$

Since $\sqrt{n}\,(\underline{\theta}_n - \underline{\theta}_\circ)$ is bounded in n, by (C3) (ii) and (iv),

$$\sup_x |\sqrt{n}\,(\underline{\theta}_n - \underline{\theta}_\circ)^t \cdot [(\partial/\partial \underline{\theta})\, G_n(x;\underline{\beta}_n',\underline{\theta}_n') - (\partial/\partial \underline{\theta})\, H(x;\underline{\beta}_\circ,\underline{\theta}_n'')]| \to 0$$

as $n \to \infty$. Finally, by (C2) (v) and (C3) (ii), $\sup_x |\epsilon_{1n}^t \cdot (\partial/\partial \underline{\theta})\, H(x;\underline{\beta}_\circ,\underline{\theta}_n^*)| \overset{p}{\to} 0$ as $n \to \infty$ and hence $\sup_x |\epsilon_{4n}(x)| \overset{p}{\to} 0$ as $n \to \infty$.

Because of the continuity properties (C3) (ii), (iii), (9), and (C3) (iv), we obtain

$$\begin{aligned} \hat{\alpha}_n(x) &= n^{-\frac{1}{2}}K(G_n(x;\underline{\beta}_n,\underline{\theta}_n),n) + \gamma^t \cdot (\partial/\partial \underline{\beta})\, G_n(x;\underline{\beta}_n,\underline{\theta}_n) \\ &\quad + w(x;\underline{\theta}_n) - \{n^{-\frac{1}{2}} \sum_{j=1}^{n} \ell_n(X_j,\underline{\beta}_n,\underline{\theta}_n) + A\cdot\gamma\}^t \cdot (\partial/\partial \underline{\theta})\, G_n(x;\underline{\beta}_n,\underline{\theta}_n) \\ &\quad + \epsilon_{5n}(x) \text{ ,} \end{aligned}$$

where $\sup_x |\epsilon_{5n}(x)| \overset{p}{\to} 0$ as $n \to \infty$.

Now, since $\sum_{j=1}^{n} \ell_n(X_j,\underline{\beta}_n,\underline{\theta}_n) = \int \ell_n(x,\underline{\beta}_n,\underline{\theta}_n)d_x\, nF_n(x)$

$= \int \ell_n(x,\underline{\beta}_n,\underline{\theta}_n)d_x\, n[F_n(x) - G_n(x;\underline{\beta}_n,\underline{\theta}_n)]$ (by (C2) (iii)), we obtain under A_n,

$$\hat{\alpha}_n(x) = Q(x,n;G_n,\underline{\beta}_n,\underline{\theta}_n) + \{\int \ell_n(x,\underline{\beta}_n,\underline{\theta}_n)d_x\, \epsilon_{2n}(x)\}^t \cdot (\partial/\partial \underline{\theta})\, G_n(x;\underline{\beta}_n,\underline{\theta}_n) + \epsilon_{5n}(x) \text{ .}$$

By Lemma 1 and (C3) (ii),

$$\sup_x \left| \{\int \ell_n(x,\underline{\beta}_n,\underline{\theta}_n)\, d_x\, \epsilon_{2n}(x)\}^t \cdot (\partial/\partial\, \underline{\beta})\, G_n(x;\underline{\beta}_n,\underline{\theta}_n) \right| \xrightarrow{a.s.} 0 \quad \text{as} \quad n \to \infty$$

and hence (6) is proved. It is clear from (7) that $\mathcal{G}(x,n;G_n,\underline{\beta}_n,\underline{\theta}_n)$ is a Gaussian process.

Proof of Corollary 1. Since $(\underline{\beta}_n,\underline{\theta}_n) \to (\underline{\beta}_\circ,\underline{\theta}_\circ)$, as $n \to \infty$, by the continuity conditions (C3) (i), (ii) and the convergence conditions (C3) (iii), (iv), we have

$$\sup_x \| (\partial/\partial (\underline{\beta},\underline{\theta})^t)\, G_n(x;\underline{\beta}_n,\underline{\theta}_n) - (\partial/\partial (\underline{\beta},\underline{\theta})^t) H(x;\underline{\beta}_\circ,\underline{\theta}_\circ) \| \to 0 ,$$

$$\sup_x |\, w(x;\underline{\theta}_n) - w(x;\underline{\theta}_\circ) \,| \to 0 ,$$

and

$$\sup_x |\, G_n(x;\underline{\beta}_n,\underline{\theta}_n) - H(x;\underline{\beta}_\circ,\underline{\theta}_\circ) \,| \to 0 , \quad \text{as} \quad n \to \infty .$$

By (C2) (vii) and (ix), $\sup_x \| \ell_n(x,\underline{\beta}_n,\underline{\theta}_n) - \ell(x,\underline{\beta}_\circ,\underline{\theta}_\circ) \| \to \underline{0}$ as $n \to \infty$. Finally, by the continuous sample path properties of the Kiefer process, the result follows.

The proof of Theorem 2 is similar to that of Theorem 1 and will be omitted.

Remark 3. The Gaussian process $\mathcal{G}(x,n;H,\underline{\beta}_\circ,\underline{\theta}_\circ)$ defined by (8) has mean

$$E\, \mathcal{G}(x,n;H,\underline{\beta}_\circ,\underline{\theta}_\circ) = -(A\cdot\gamma)^t\, (\partial/\partial\, \underline{\theta})\, H(x;\underline{\beta}_\circ,\underline{\theta}_\circ)$$
$$+ \gamma^t \cdot (\partial/\partial\, \underline{\beta})\, H(x;\underline{\beta}_\circ,\underline{\theta}_\circ) + w(x;\underline{\theta}_\circ)$$

and covariance function

$$E\, \mathcal{G}(x,n;H,\underline{\beta}_\circ,\underline{\theta}_\circ)\, \mathcal{G}(y,m;H,\underline{\beta}_\circ,\underline{\theta}_\circ)$$
$$= \{(n \wedge m)/\sqrt{nm}\} \cdot \{H(x;\underline{\beta}_\circ,\underline{\theta}_\circ) \wedge H(y;\underline{\beta}_\circ,\underline{\theta}_\circ) - H(x;\underline{\beta}_\circ,\underline{\theta}_\circ)\, H(y;\underline{\beta}_\circ,\underline{\theta}_\circ)$$
$$- J(x;\underline{\beta}_\circ,\underline{\theta}_\circ)^t \cdot (\partial/\partial\, \underline{\theta})\, H(y;\underline{\beta}_\circ,\underline{\theta}_\circ) - J(y;\underline{\beta}_\circ,\underline{\theta}_\circ)^t \cdot (\partial/\partial\, \underline{\theta})\, H(x;\underline{\beta}_\circ,\underline{\theta}_\circ)$$
$$+ (\partial/\partial\, \underline{\theta})\, H(x;\underline{\beta}_\circ,\underline{\theta}_\circ)^t \cdot L(\underline{\theta}_\circ) \cdot (\partial/\partial\, \underline{\theta})\, H(y;\underline{\beta}_\circ,\underline{\theta}_\circ)\} ,$$

where $J(x;\underline{\beta}_\circ,\underline{\theta}_\circ) = \int_{-\infty}^{x} \ell(z,\underline{\beta}_\circ,\underline{\theta}_\circ)\, d\, H(z,\underline{\beta}_\circ,\underline{\theta}_\circ)$. We have for each n ,

$G(x,n;H,\underline{\beta}_\circ,\underline{\theta}_\circ) \stackrel{\mathcal{D}}{=} G(x,1;H,\underline{\beta}_\circ,\underline{\theta}_\circ)$, and since $K(t,1) \stackrel{\mathcal{D}}{=} B(t)$ a Brownian Bridge, under $\{A_n\}$

$$\hat{\alpha}_n(x) \stackrel{\mathcal{D}}{\to} B(H(x;\underline{\beta}_\circ,\underline{\theta}_\circ)) - \{\int \ell(x,\underline{\beta}_\circ,\underline{\theta}_\circ)\, d_x\, B(H(x;\underline{\beta}_\circ,\underline{\theta}_\circ))\}^t \cdot$$

$$(\partial/\partial\, \underline{\theta})\, H(x;\underline{\beta}_\circ,\underline{\theta}_\circ)$$

$$- (A\cdot\gamma)^t \cdot (\partial/\partial\, \underline{\theta})\, H(x;\underline{\beta}_\circ,\underline{\theta}_\circ) + \gamma^t (\partial/\partial\, \underline{\beta})\, H(x;\underline{\beta}_\circ,\underline{\theta}_\circ) + w(x;\underline{\theta}_\circ) .$$

It is clear that the Gaussian processes defined by (7) and (8) depend on $\underline{\theta}_n$ and $\underline{\theta}_\circ$, the true unknown values of $\underline{\theta}$. Thus, in general, Theorem 2 and Corollary 1 cannot be applied to test the composite hypotheses (1) and (3). However, we can remedy the situation as follows: let $G(x,n;H,\underline{\beta}_\circ,\hat{\underline{\theta}}_n)$ be defined by (8) with $\underline{\theta}_\circ$ replaced by $\hat{\underline{\theta}}_n$, the estimator of $\underline{\theta}$. Then, we have

<u>Theorem 3</u>. Under the conditions of Theorem 1,

$$\sup_x |\, G(x,n;H,\underline{\beta}_\circ,\theta_\circ) - G(x;n;H,\underline{\beta}_\circ,\hat{\underline{\theta}}_n)|\, \stackrel{P}{\to} 0 \quad \text{as} \quad n \to \infty$$

and hence, under $\{A_n\}$,

$$\sup_x |\, \hat{\alpha}_n(x) - G(x,n;H,\underline{\beta}_\circ,\hat{\underline{\theta}}_n)\,|\, \stackrel{P}{\to} 0 , \quad \text{as} \quad n \to \infty .$$

If we assume that the conditions of Theorem 2 are satisfied, then under the null hypothesis $H_\circ$,

$$\sup_x |\, \hat{\alpha}_n(x) - \hat{G}(x,n)\,|\, \stackrel{P}{\to} 0 \quad \text{as} \quad n \to \infty ,$$

where $\hat{G}(x,n)$ is defined by (8) with $\underline{\theta}_\circ$ replaced by $\hat{\underline{\theta}}_n$ and with γ and $w(x;\underline{\theta}_\circ)$ equal to zero.

The proof of Theorem 3 follows from the fact that $(\hat{\underline{\theta}}_n - \underline{\theta}_\circ) \stackrel{P}{\to} \underline{0}$, as $n \to \infty$, and the continuity properties (uniformly in x) of the functions and the Kiefer process involved in $G(x,n;H,\underline{\beta}_\circ,\underline{\theta})$.

Remark 4. Similar statements to our Theorems 1, 2, and 3 and Corollary 1 can be made using the sequence of Brownian Bridges of Theorem A, instead of the Kiefer process. However, our approach of working with Gaussian processes defined in terms of Kiefer processes also gives one the joint distribution of $\mathcal{G}$ in x and in n, while a corresponding construction in terms of Brownian Bridges $B_n(\cdot)$ would only give a Gaussian process $\mathcal{G}_n$ in x for each n.

3. The Maximum Likelihood Estimation Case.

As mentioned in Example 1, maximum likelihood estimators often satisfy conditions (C1) and (C2). Suppose $\{\hat{\underline{\theta}}_n\}$ is a sequence of maximum likelihood estimators satisfying conditions (C2), with $\ell(x,\underline{\beta}_0,\underline{\theta}_0) =$ $(\partial/\partial\,\underline{\theta}) \log h(x;\underline{\beta}_0,\underline{\theta}_0)^t \cdot I^{-1}(\underline{\theta}_0)$ and $\ell_n(x,\underline{\beta}_n,\theta_n) = (\partial/\partial\,\underline{\theta}) \log g_n(x;\underline{\beta}_n,\underline{\theta}_n)^t \cdot I_n^{-1}(\underline{\theta}_n)$, where $I(\underline{\theta})$, respectively $I_n(\underline{\theta})$, is the Fisher information matrix with respect to h, respectively g_n, and I^{-1} is the inverse matrix of I. If we also assume (C3), then the statements of Theorem 1 and Corollary 1 are valid and $\mathcal{G}(x,n;H,\underline{\beta}_0,\underline{\theta}_0)$ is defined by

$$
\begin{aligned}
\mathcal{G}(x,n;H,\underline{\beta}_0,\underline{\theta}_0) &= n^{-\frac{1}{2}} K(H(x;\underline{\beta}_0,\underline{\theta}_0), n) \\
&\quad - \left\{\int (\partial/\partial\,\underline{\theta}) \log h(x;\underline{\beta}_0,\underline{\theta}_0)\, d_x\, n^{-\frac{1}{2}} K(H(x;\underline{\beta}_0,\underline{\theta}_0),n)\right\}^t \cdot I^{-1}(\underline{\theta}_0) \cdot \\
&\quad \cdot (\partial/\partial\,\underline{\theta})\, H(x;\underline{\beta}_0,\underline{\theta}_0) \\
&\quad - (I^{-1}(\underline{\theta}_0)\cdot I_{21}\cdot\gamma)^t \cdot (\partial/\partial\,\underline{\theta})\, H(x;\underline{\beta}_0,\underline{\theta}_0) \\
&\quad + \gamma^t \cdot (\partial/\partial\,\underline{\beta})\, H(x;\underline{\beta}_0,\underline{\theta}_0) + w(x;\underline{\theta}_0) .
\end{aligned} \tag{10}
$$

Remark 5. In this maximum likelihood case, the covariance of $\mathcal{G}$, defined by (10), simplifies to

$$
\begin{aligned}
&E\, \mathcal{G}(x,n;H,\underline{\beta}_0,\underline{\theta}_0)\, \mathcal{G}(y,m;H,\underline{\beta}_0,\underline{\theta}_0) = \\
&\quad ((n \wedge m)/\sqrt{nm}) \cdot \{H(x;\underline{\beta}_0,\underline{\theta}_0) \wedge H(y;\underline{\beta}_0,\underline{\theta}_0) - H(x;\underline{\beta}_0,\underline{\theta}_0)\cdot H(y;\underline{\beta}_0,\underline{\theta}_0) \\
&\quad - (\partial/\partial\,\underline{\theta})\, H(x;\underline{\beta}_0,\underline{\theta}_0)^t \cdot I^{-1}(\underline{\theta}_0) \cdot (\partial/\partial\,\underline{\theta})\, H(y;\underline{\beta}_0,\underline{\theta}_0) ,
\end{aligned}
$$

since $J(x,\underline{\beta}_\circ,\underline{\theta}_\circ) = (\partial/\partial\,\underline{\theta})\, H(x;\underline{\beta}_\circ,\underline{\theta}_\circ)\cdot I^{-1}(\underline{\theta}_\circ)$, assuming that

$(\partial/\partial\,\underline{\theta}) \int_{-\infty}^{x} h(x;\underline{\beta},\underline{\theta})\, dz = \int_{-\infty}^{x} (\partial/\partial\,\underline{\theta})\, h(z;\underline{\beta},\underline{\theta})\, dz$.

The limiting Gaussian process $\mathcal{G}_\circ(x,n)$, under the null hypothesis $H_\circ$ is given by (10) with γ and $w(x;\underline{\theta}_\circ)$ equal to zero. The covariance of $\mathcal{G}_\circ(x,n)$ is given in Remark 5. Our Theorem 3 of Section 2 can also be stated for maximum likelihood estimators with $\mathcal{G}$ defined by (10).

In [8] and in his Oberwolfach Conference contribution, J. Durbin has observed that if the maximum likelihood estimator of $\underline{\theta}$ is based on a randomly chosen half of the sample, the resulting estimated empirical process converges in distribution to a Brownian Bridge process. The Kolmogorov-Smirnov-type statistics based on this process become asymptotically distribution-free. This line of thinking was begun by K.C. Rao [10]. We shall prove this result here using the strong approximation methodology. A strong approximation version of Theorem 4, which follows, was proved by Csörgő, Komlós, Major, Révész and Tusnády [4] for univariate x and one-dimensional θ . We shall extend the proof in [4] in terms of weak approximation and multidimensional $\underline{\theta}$.

Let $\underline{\bar{\theta}}_n$ be the maximum likelihood estimator of $\underline{\theta}$ based on a randomly chosen half of the sample $X_1,X_2,\ldots,X_n$. Without loss of generality we can assume that it is the first half: $X_1,X_2,\ldots,X_{[n/2]}$.

<u>Theorem 5</u>. Suppose that conditions (C1) and (C3) are satisfied for maximum likelihood estimators. Then, under the null hypothesis $H_\circ$ of (1), one can construct a Kiefer process $\bar{K}(\cdot,\cdot)$ such that

$$\sup_x \left| \sqrt{n}\, [F_n(x) - H(x;\underline{\beta}_\circ,\underline{\bar{\theta}}_n)] - n^{-\frac{1}{2}}\, \bar{K}(H(x;\underline{\beta}_\circ,\underline{\theta}_\circ),\, n) \right| \xrightarrow{P} 0$$

as $n \to \infty$, and hence

$$\sqrt{n}\, [F_n(x) - H(x;\underline{\beta}_\circ,\underline{\bar{\theta}}_n)] \xrightarrow{\mathcal{D}} B(H(x;\underline{\beta}_\circ,\underline{\theta}_\circ)) .$$

Proof. Let $F_n^{(1)}(x)$, resp. $F_n^{(2)}(x)$ be the empirical d.f. based on the sample $X_1, X_2, \dots, X_{[n/2]}$, resp. $X_{[n/2]+1}, \dots, X_n$ and let $K_1(\cdot,\cdot)$ and $K_2(\cdot,\cdot)$ be the Kiefer processes of Theorem A for which

$$\sup_x | (n/2) [F_n^{(1)}(x) - H(x;\underline{\beta}_\circ,\underline{\theta}_\circ)] - K_1(H(x;\underline{\beta}_\circ,\underline{\theta}_\circ), n/2) |$$

$$\overset{a.s.}{=} \Theta\{\log^2(n/2)\}$$

and

$$\sup_x | (n/2) [F_n^{(2)}(x) - H(x;\underline{\beta}_\circ,\underline{\theta}_\circ)] - K_2(H(x;\underline{\beta}_\circ,\underline{\theta}_\circ), n/2) |$$

$$\overset{a.s.}{=} \Theta\{\log^2(n/2)\} .$$

Without loss of generality, we can assume that $K_2(\cdot,\cdot)$ is independent from $K_1(\cdot,\cdot)$ and also from the sample $X_1, X_2, \dots, X_{[n/2]}$. We obtain

$$n[F_n(x) - H(x;\underline{\beta}_\circ,\bar{\underline{\theta}}_n)] = (n/2) [F_n^{(1)}(x) - H(x;\underline{\beta}_\circ,\underline{\theta}_\circ)]$$

$$+ (n/2) [F_n^{(2)}(x) - H(x;\underline{\beta}_\circ,\underline{\theta}_\circ)] - n[H(x;\underline{\beta}_\circ,\bar{\underline{\theta}}_n) - H(x;\underline{\beta}_\circ,\underline{\theta}_\circ)]$$

$$= K_1(H(x;\underline{\beta}_\circ,\underline{\theta}_\circ), n/2) + K_2(H(x;\underline{\beta}_\circ,\underline{\theta}_\circ), n/2)$$

$$- n[H(x;\underline{\beta}_\circ,\bar{\underline{\theta}}_n) - H(x;\underline{\beta}_\circ,\underline{\theta}_\circ)] + \varepsilon_{6n}(x) ,$$

where $\sup_x | \varepsilon_{6n}(x) | \overset{a.s.}{=} \Theta\{\log^2(n/2)\}$. As in the proof of Theorem 1, we obtain

$$n[F_n(x) - H(x;\underline{\beta}_\circ,\bar{\underline{\theta}}_n)] = K_1(H(x;\underline{\beta}_\circ,\underline{\theta}_\circ), n/2)$$

(11) $$+ K_2(H(x;\underline{\beta}_\circ,\underline{\theta}_\circ), n/2)$$

$$- 2\{\int (\partial/\partial\, \underline{\theta}) \log h(x;\underline{\beta}_\circ,\underline{\theta}_\circ)\, d_x K_1(H(x;\underline{\beta}_\circ,\underline{\theta}_\circ), n/2)\}^t$$

$$\cdot I^{-1}(\underline{\theta}_\circ) \cdot (\partial/\partial\, \underline{\theta})\, H(x;\underline{\beta}_\circ,\underline{\theta}_\circ) + \varepsilon_{7n}(x) ,$$

where $\sup_x | n^{-\frac{1}{2}} \varepsilon_{7n}(x) | \overset{p}{\to} 0$ as $n \to \infty$.

The process $\bar{K}(H(x;\underline{\beta}_\circ,\underline{\theta}_\circ),n) = K_1(H(x;\underline{\beta}_\circ,\underline{\theta}_\circ),n/2)$

$$+ K_2(H(x;\underline{\beta}_\circ,\underline{\theta}_\circ),n/2) - 2\{\int (\partial/\partial\,\underline{\theta}) \log h(x;\underline{\beta}_\circ,\underline{\theta}_\circ)\, d_x\, K_1(H(x;\underline{\beta}_\circ,\underline{\theta}_\circ),n/2)\}^t$$

$$\cdot I^{-1}(\underline{\theta}_\circ)\cdot(\partial/\partial\,\underline{\theta})\, H(x;\underline{\beta}_\circ,\underline{\theta}_\circ)$$

is clearly a Gaussian Process with mean zero and, by calculation, $\bar{K}$ has covariance $E\,\bar{K}(H(x;\underline{\beta}_\circ,\underline{\theta}_\circ),n)\cdot\bar{K}(H(y;\underline{\beta}_\circ,\underline{\theta}_\circ),m)$

$$= (n \wedge m)\,\{H(x;\underline{\beta}_\circ,\underline{\theta}_\circ) \wedge H(y;\underline{\beta}_\circ,\underline{\theta}_\circ) - H(x;\underline{\beta}_\circ,\underline{\theta}_\circ)\, H(y;\underline{\beta}_\circ,\underline{\theta}_\circ)\}\ .$$

Hence $\bar{K}(\cdot,\cdot)$ is a Kiefer process and on dividing (11) through by $\sqrt{n}$, the result follows.

Remark 6. As observed by Durbin in [8], the asymptotic distribution-freeness of the Kolmogorov-Smirnov-type statistics, based on the empirical process when unknown parameters are estimated from a randomly chosen half of the sample, thus gained (through Theorem 5), is illusionary. A randomization has been introduced; the value of $\bar{\underline{\theta}}_n$ depends on the particular half-sample chosen. With this randomization, the empirical process of Theorem 5 behaves as if the unknown parameters were completely specified.

Remark 7. The results and methodology of Theorem 1 and Corollary 1 can be used to study the asymptotic power of many statistics based on the estimated empirical process against various classes of alternatives. Such a study was conducted in Burke [3] for statistics based on the one and several sample quantile and multivariate empirical processes.

A multivariate version of Theorem A has been proved by Csörgő and Révész [5]. The results of this paper are extended for the estimated multivariate empirical process in Burke and Csörgő [1]. In the case of maximum likelihood estimators, strong approximations (with rates) can be found in Csörgő, Komlós, Major, Révész, and Tusnády [4], for the estimated univariate empirical process with one-dimensional parameter space, and in Burke and Csörgő [2] for the estimated multivariate empirical process with multidimensional parameter space. Also, strong approximations of the univariate space quantile process have been proved by Csörgő and Révész [6]. For weak and strong approximations of the quantile process when parameters are estimated, we refer to Burke and Csörgő [1], [2].

References

[1] Burke, M.D. and Csörgő, M.: Weak approximations of the estimated sample quantile and multivariate empirical processes via strong approximation methods. (To appear).

[2] Burke, M.D. and Csörgő, M.: Strong approximations of the estimated quantile and multivariate empirical processes. (To appear).

[3] Burke, M.D.: On the asymptotic power of some statistics based on the one and several sample quantile and multivariate empirical processes. (To appear).

[4] Csörgő, M., Komlós, J., Major, P., Révész, P., and Tusnády, G.: On the empirical process when parameters are estimated. Proceedings of the Seventh Prague Conference, 1974. (To appear).

[5] Csörgő, M. and Révész, P.: A strong approximation of the multivariate empirical process. Studia Sci. Math. Hungar. (To appear).

[6] Csörgő, M. and Révész, P.: Strong approximations of the quantile process. Ann. Statist. (To appear).

[7] Durbin, J.: Weak convergence of the sample distribution function when parameters are estimated. Ann. Statist. 1 279-290 (1973).

[8] Durbin, J.: Distribution theory for tests based on the sample distribution function. Regional Conference Series in Applied Math. 9 SIAM, Philadelphia (1973).

[9] Komlós, J., Major, P. and Tusnády, G.: An approximation of partial sums of independent R.V.'s and the sample DF. I. Z. Wahrscheinlichkeitstheorie verw. Gebiete 32 111-131 (1975).

[10] Rao, K.C.: The Kolmogoroff, Cramér-von Mises, chi-square statistics for goodness-of-fit tests in the parametric case. Abstract 133-6. Bulletin IMS 1 87 (1972).

M. Csörgő, M.D. Burke,
Department of Mathematics,
Carleton University,
OTTAWA, Canada. K1S 5B6

ON THE ERDŐS-RÉNYI INCREMENTS AND THE P. LÉVY MODULUS OF CONTINUITY OF A KIEFER PROCESS

Miklós Csörgő[1] and Arthur H.C. Chan[2]
Carleton University, Ottawa
Canada K1S 5B6

1. Introduction and Summary of Some Recent Strong Embeddings of the Empirical Process.

Multi-time parameter Wiener and Kiefer processes play a prominant role in a number of recent strong invariance principles for the multivariate empirical process.

Definition 1: A d-parameter Wiener process $W(\underline{x})$, $\underline{x} \in [0,\infty)^d$, defined on some probability space (Ω, A, P), is a separable Gaussian satisfying the following conditions:

(i) $W(\cdot)$ is almost surely (a.s.) continuous on $[0,\infty)^d$,

(ii) $W(\underline{x}) \overset{a.s.}{=} 0$ if $|\underline{x}| = x_1 x_2 \ldots x_d$ is zero and if $|\underline{x}| > 0$, then

$P\{W(\underline{x}) \le \lambda\} = \Phi(\lambda|\underline{x}|^{-\frac{1}{2}})$, $-\infty < \lambda < +\infty$, where $\Phi(u) = \frac{1}{(2\pi)^{\frac{1}{2}}} \int_{-\infty}^{u} e^{-t^2/2}\, dt$,

(iii) $W(\underline{x})$ has independent increments over disjoint intervals.

Definition 2: A (d+1)-parameter Kiefer process $K(\underline{x},y)$, $\underline{x} \in I^d$, $I = [0,1]$, $y \ge 0$ is defined as

$$K(\underline{x},y) = W(\underline{x},y) - |\underline{x}|\, W(\underline{1},y),$$

where $W(\underline{x},y)$ is a (d+1)-parameter Wiener process.

Consequently, the covariance function of a Kiefer process is

$$EK(\underline{x}_1,y_1)K(\underline{x}_2,y_2) = (y_1 \wedge y_2)(|\underline{x}_1 \wedge \underline{x}_2| - |\underline{x}_1||\underline{x}_2|),$$

where the operation $\wedge$ stands for taking minimum componentwise when applied to higher dimensional vectors.

---

[1] Research partially supported by a Canadian N.R.C. Grant.

[2] Research supported by a Canadian N.R.C. Post-graduate Scholarship.

Definition 3: A d-parameter Brownian bridge $B(\underline{x})$, $\underline{x} \in I^d$, is defined as

$$B(\underline{x}) = W(\underline{x}) - |\underline{x}|W(\underline{1}) ,$$

where $W(\underline{x})$ is a d-dimensional Wiener process.

Let $\underline{X}_1, \underline{X}_2, \ldots$ be a sequence of independent random variables uniformly distributed over the unit cube I^d of the d-dimensional Euclidean space R^d. Let $F_n(\cdot)$ be the empirical distribution based on the random sample $\underline{X}_1, \underline{X}_2, \ldots, \underline{X}_n$ and let

$$\alpha_n(\underline{x}) = n^{\frac{1}{2}}(F_n(\underline{x}) - |\underline{x}|) , \ \underline{x} \in I^d ,$$

be the empirical process.

As to strong approximations of this empirical process of uniformly distributed random variables by a sequence of Brownian bridges resp. Kiefer process, the following is known:

Theorem A: One can construct a probability space (Ω, A, P) with a sequence of independent random variables uniformly distributed over I^d, a sequence of Brownian Bridges $\{B_n(\cdot)\}$ and a Kiefer process $K(\cdot,\cdot)$ such that

$$\sup_{\underline{x}} |\alpha_n(\underline{x}) - B_n(\underline{x})| \overset{a.s.}{=} O(r_1(n))$$

$$\sup_{\underline{x}} |n^{\frac{1}{2}} \alpha_n(\underline{x}) - K(\underline{x}, n)| \overset{a.s.}{=} O(r_2(n)) ,$$

with $r_1(n) = n^{-\frac{1}{2}} \log n$ and $r_2(n) = \log^2 n$ if $d = 1$ (Komlós, Major, Tusnády, 1975), and with $r_1(n) = n^{-\frac{1}{2(d+1)}}(\log n)^{3/2}$ and $r_2(n) = n^{\frac{d+1}{2(d+2)}} \log^2 n$ for any $d = 1,2,\ldots$ (Csörgő, Révész, 1975).

Kiefer (1972) was the first one to prove a strong embedding theorem for the one dimensional empirical process ($d = 1$ and $r_2(n) = n^{1/3}(\log n)^{2/3}$ in Theorem A). Müller (1970) proved the weak convergence of $\alpha_n(x)$, $x \in I$, to a Gaussian process over $I \times [0,\infty)$ with covariance function as that of $K(x,n)/n^{\frac{1}{2}}$ of Definition 2. The assumption that we have a uniformly distributed random sample is no restriction when in Theorem A as long as $d = 1$. A generalization of Theorem A to higher dimensions with arbitrary continuous distribution functions is not so immediate (cf. Csörgő, Révész, 1976a). In his recent paper in these Lecture Notes Révész

(1976) studies some properties of the stochastic set function $\alpha_n(A) = \int_A d\,\alpha_n(\underline{x})$, when A runs over the class of Borel subsets of I^d, having d-times differentiable boundaries. He proves a large deviation theorem similar to that of Kiefer (1961) and a law of iterated logarithm for $\sup \alpha_n(A)$ and also a strong invariance principle for $\alpha_n(A)$.

The aim of our present paper is to study how large the increments of the multi-time parameter Wiener and Kiefer processes can be. In the next section we summarize some recent results of Chan (1976) and Csörgő and Révész (1976b), which concern the Erdős-Rényi type increments of the Wiener process and its P. Lévy modulus of continuity. Then in Sections 3 and 4, we prove similar results for the Kiefer process. The results of these latter sections are believed to be new and the proofs will not be published elsewhere.

2. On the Erdős-Rényi Increments and the P. Lévy Modulus of Continuity of Multi-Time Wiener Processes.

For each fixed $\underline{x} \in [0,\infty)^d$, define $H_{\underline{x}}^{(i)} : [0,\infty)^d \to [0,\infty)^d$ by

$$H_{\underline{x}}^{(i)}(\underline{y}) = (y_1,\ldots,y_{i-1},x_i,y_{i+1},\ldots,y_d), \quad 1 \leq i \leq d .$$

For $\underline{x} \leq \underline{y}$ we define

$$\langle \underline{x},\underline{y} \rangle = \{\underline{z} \in [0,\infty)^d : \underline{x} \leq \underline{z} \leq \underline{y}\}, \quad \langle \underline{x} \rangle = \langle \underline{0},\underline{x} \rangle \quad \text{and}$$

$$\mathrm{int}\langle \underline{x},\underline{y} \rangle = \{\underline{z} \in [0,\infty)^d : \underline{x} < \underline{z} < \underline{y}\} .$$

Suppose that $\underline{x} < \underline{y}$. Define

$$\begin{aligned} W(\underline{x},\underline{y}) = W(\underline{y}) &+ (-1) \sum_{i=1}^{d} W(H_{\underline{x}}^{(i)}(\underline{y})) \\ &+ (-1)^2 \sum_{i<j} W(H_{\underline{x}}^{(i)} H_{\underline{x}}^{(j)}(\underline{y})) + \ldots \\ &+ (-1)^{d-1} \sum_{i} W(H_{\underline{y}}^{(i)}(\underline{x})) \\ &+ (-1)^d W(\underline{x}) . \end{aligned}$$

Using this latter definition, property (iii) of $W(\underline{x})$ of Definition 1 of Section 1 can be described via saying that $W(\underline{x},\underline{y})$ and $W(\underline{x}',\underline{y}')$ are independent of $\mathrm{int}\langle\underline{x},\underline{y}\rangle \cap \mathrm{int}\langle\underline{x}',\underline{y}'\rangle = \emptyset$. We also have

$$P\{W(\underline{x},\underline{y}) < \lambda\} = \Phi(\lambda|\underline{y}-\underline{x}|^{-\frac{1}{2}}), \quad -\infty < \lambda < +\infty .$$

When $d = 1$, the Erdős-Rényi (1970) law of large numbers has the following special form in the Gaussian case.

Theorem B: (Erdős, Rényi 1970). Let $W(x)$, $x \geq 0$, be a standard Brownian motion. Denote by $[y]$ the greatest integer less than y. Then, for each $c > 0$, we have

$$\lim_{N\to\infty} \max_{0\leq j\leq N-[c \log N]} \frac{W(j+[c \log N]) - W(j)}{[c \log N]} \overset{a.s.}{=} \sqrt{\frac{2}{c}} .$$

Also, in their just quoted paper, it is mentioned that Theorem B yields

$$\lim_{h\to 0} P\{|\sup_{0\leq x\leq 1-h} \frac{W(x+h) - W(x)}{\sqrt{2h \log h^{-1}}} - 1| > \epsilon\} = 0$$

for each $\epsilon > 0$. This, of course, is the "in probability" version of P. Lévy's modulus of continuity for Brownian motion:

Theorem C: (P. Lévy 1937). Let $W(x)$, $0 \leq x \leq 1$, be a Brownian motion on $[0,1]$. Then

$$\overline{\lim_{h\to\infty}} \sup_{0\leq x\leq 1-h} \frac{W(x+h) - W(x)}{\sqrt{2h \log h^{-1}}} \overset{a.s.}{=} 1 .$$

Theorem C is actually true with lim instead of $\overline{\lim}$ (cf. Orey and Taylor 1974). Also, Orey and Pruitt (1973) have extended Theorem C to the case of the d-parameter Wiener process.

Let us reserve the symbols $\underline{i}$, $\underline{j}$ and $\underline{k}$ for points in $[0,\infty)^d$ having integer coordinates.

The following is a d-dimensional generalization of Theorem B.

Theorem D: (Chan, 1976). Let $\underline{a}_N \in [0,\infty)^d$, $N = 1,2,\ldots$. Suppose that $\underline{a}_N$ satisfies the following condition

(2.1) $$\lim_{N\to\infty} \frac{|a_N|}{N^\delta} = 0 \text{ , for each } \delta > 0 .$$

Then, for each $c > 0$,

(2.2) $$\lim_{N\to\infty} \max_{\underline{0}\le\underline{j}\le\underline{N}-\underline{a}_N} \frac{W(\underline{j}, \underline{j}+\underline{a}_N)}{\sqrt{|\underline{a}_N|[c \log N]}} \overset{a.s.}{=} \sqrt{\frac{2d}{c}} ,$$

where $\underline{N} = (N,\ldots,N)$.

Corollary D1: (Chan, 1976). Suppose $\{\underline{a}_N\}$ satisfies

(2.3) $$\lim_{N\to\infty} \frac{|\underline{a}_N|}{[\log N]} = c > 0 .$$

Then we have

(2.4) $$\lim_{N\to\infty} \max_{\underline{0}\le\underline{j}\le\underline{N}-\underline{a}_N} \frac{W(\underline{j}, \underline{j}+\underline{a}_N)}{|a_N|} \overset{a.s.}{=} \sqrt{\frac{2d}{c}} .$$

In particular, when $d = 1$ and $a_N = [c \log N]$, we have Theorem B, on observing that definition of $W(\underline{x},\underline{y})$ implies $W(j+k) - W(j) = W(j,j+k)$. Also, in connection with (2.4) we observe that $E((W(\underline{j},\underline{j} + \underline{a}_N))^2) = |\underline{a}_N|$.

Remark 1: Suppose that $c = 0$ or $c = \infty$ in (2.3). Then (2.4) diverges if $c = 0$ and converges to 0 if $c = \infty$. Therefore, $\log N$ is the best rate in (2.4), resulting in the limit $\sqrt{\frac{2d}{c}}$, the Erdős-Rényi characteristic number of $W(\underline{x})$, $\underline{x} \in [0,\infty)^d$.

In the case of $d = 1$, if we let $a_N = 1$, then we have

$$\max_{0\le j\le N-1} \frac{W(j,j+1)}{\sqrt{[\log N]}} \overset{a.s.}{\to} \sqrt{2} , \text{ as } N \to \infty .$$

Thus we proved

Corollary D2: (Chan, 1976). Let $X_1, X_2, \ldots$ be a sequence of i.i.d. normal r.v. with mean zero and variance one. Then we have

(2.5) $$\lim_{N\to\infty} \max_{1\le i\le N} \frac{X_i}{\sqrt{2[\log N]}} \overset{a.s.}{=} 1 .$$

The next theorem is a stronger version of Theorem D, and was proved with the aim of trying to link the Erdős-Rényi and the P. Lévy types of modulus of continuity together. (Anticipating a strong connection between them, sometimes we call the results of Theorems B and D Erdős-Rényi type modulae of continuity instead of the more natural Erdős-Rényi increment terminology.)

Theorem E: (Chan, 1976). Let $\{\underline{a}_N\} \in [0,\infty)^d$ be such that the coordinates of each $\underline{a}_N$ are positive integers. Assume that the sequence $\{\underline{a}_N\}$ satisfies (2.1) and, letting $\underline{a}_N = (a_{1N},\ldots,a_{dN})$, assume also

$$\lim_{N\to\infty} a_{iN} = \infty\ ,\quad 1 \le i \le N\ . \tag{2.6}$$

Define, for each N and $\underline{k} \le \underline{N}$,

$$S(N,k) = \sup_{\underline{0}\le\underline{y}\le\underline{k}}\ \sup_{\underline{0}\le\underline{x}\le\underline{N}-\underline{y}} \frac{W(\underline{x},\underline{x}+\underline{y})}{\sqrt{2d|\underline{k}|[\log N]}}\ . \tag{2.7}$$

Then, for each $\epsilon > 0$, there is a $\gamma = \gamma(\epsilon) > 0$ such that

$$P\{S(N,\underline{a}_N) > 1+\epsilon\} \le AN^{p\gamma}\ . \tag{2.8}$$

Furthermore,

$$\lim_{N\to\infty} S(N,\underline{a}_N) \overset{a.s.}{=} 1\ . \tag{2.9}$$

Corollary E1: (Chan, 1976). Let $\{\underline{a}_N\}$ be as in Theorem E. Then

$$\lim_{N\to\infty}\ \sup_{\underline{0}\le\underline{x}\le\underline{N}-\underline{a}_N} \frac{W(\underline{x},\underline{x}+\underline{a}_N)}{\sqrt{2d|\underline{a}_N|[\log N]}} \overset{a.s.}{=} 1\ . \tag{2.10}$$

In particular, when $d = 1$ and $a_N = [c \log N]$, for each $c > 0$

$$\lim_{N\to\infty}\ \sup_{\underline{0}\le\underline{x}\le N-[c\log N]} \frac{W(x+[c\log N])-W(x)}{[c\log N]} \overset{a.s.}{=} \sqrt{\frac{2}{c}}\ . \tag{2.11}$$

The statement of (2.10), resp. that of (2.11), is a stronger version of (2.2), resp. that of Theorem B.

As we have already mentioned earlier in this section, Erdős and Rényi (1970) noticed that Theorem B implies a weaker form of Theorem C. It does not, however, seem to be obvious that Theorem B should imply Theorem C. Indeed, if Theorem B is formulated as it is, then we were unable to prove this latter implication.

Actually, the duality of Theorems B and C follows from Theorem E, that is both of them are implied by the latter. Corollary E1 above (cf. (2.11) and Theorem B) is the easier part of our statement. As to the problem of Theorem E implying Theorem C, we refer to Section 4 of Chan's paper (1976).

Orey and Pruitt (1973) give the following extension of Theorem C to the case when $d > 1$.

<u>Theorem F</u>: (Orey, Pruitt, 1973). Let J denote an interval and let $I = \langle \underline{0}, \underline{1} \rangle$. Then

$$\lim_{h\to 0} \sup_{\substack{J\subset I \\ |J|\le h}} \frac{W(J)}{\sqrt{2|J|\log|J|^{-1}}} \overset{a.s.}{=} 1, \tag{2.12}$$

where $|J| = |\underline{y} - \underline{x}|$ if $J = \langle \underline{x}, \underline{y} \rangle$.

It is natural to ask if one can prove (2.12) from Theorem E. Then answer is "not quite", and this is mainly because one lacks total ordering in $[0,\infty)^d$. One can show, however, that

$$\overline{\lim_{h\to 0}} \sup_{\underline{0}\le\underline{x}\le\underline{1}-\underline{h}} \frac{W(\underline{x}, \underline{x}+\underline{h})}{\sqrt{2|\underline{h}|\log|\underline{h}|^{-1}}} \overset{a.s.}{=} 1, \tag{2.13}$$

where $\underline{h} = (h, h, \ldots, h) > \underline{0}$.

This latter statement is obviously weaker than (2.12). Judging from the case when $d = 1$, it is inviting to conjecture that, when $d > 1$, one should be able to prove Theorem F from a Theorem E-type result.

More recently Csörgő and Révész (1976b) studied the increments of a one-time parameter Wiener process on subintervals of length $a_T \le T$ of $[0,T]$. They prove

<u>Theorem G</u>: (Csörgő, Révész, 1976b). Let $a_T\,(T\ge 0)$ be a monotonically non-decreasing continuous function of T for which

(i) $0 < a_T \le T \ (T \ge 0)$,

(ii) $a_{\theta^{k+1}}/a_{\theta^k} \le \theta$ for any $\theta > 1$ if k is big enough,

(iii) T/a_T is monotonically non-decreasing. Then

$$\overline{\lim_{T\to\infty}} \sup_{0\le t\le T-a_T} \beta_T|W(t+a_T) - W(t)| \overset{a.s.}{=} 1 \tag{2.14}$$

and

$$(2.15) \qquad \overline{\lim_{T\to\infty}} \sup_{0\le t\le T-a_T} \sup_{0\le s\le a_T} \beta_T |W(t+s) - W(t)| \overset{a.s.}{=} 1 ,$$

where $\beta_T = (2a_T[\log \frac{T}{a_T} + \log \log T])^{-\frac{1}{2}}$.

If we also have

$$(iv) \qquad \sum_{n=1}^{\infty} \exp\{-(\frac{n}{a_n})^{\epsilon} (\frac{1}{\log n})^{1-\epsilon}\} < \infty$$

for any $0 < \epsilon < 1$, then

$$(2.16) \qquad \lim_{T\to\infty} \sup_{0\le t\le T-a_T} \beta_T |W(t+a_T) - W(t)| \overset{a.s.}{=} 1 ,$$

and

$$(2.17) \qquad \lim_{T\to\infty} \sup_{0\le t\le T-a_T} \sup_{0\le s\le a_T} \beta_T |W(t+s) - W(t)| \overset{a.s.}{=} 1 .$$

Theorem G, in turn, is used by Csörgő and Révész (1976) to study similar increments of partial sum processes via the recent strong invariance principle of Komlós, Major and Tusnády (1976) and that of Major (1976).

A compact theorem like Theorem G is also possible for the multi-time parameter Wiener process (cf. Chan, Csörgő and Révész, 1976).

The proofs of theorems stated in this section will be published elsewhere. As to the nature of these proofs, those of Section 4 are indicative.

3. On the Increments of a Kiefer Process.

Let $K(\underline{x},t)$, $\underline{x} = (x_1,\ldots,x_d) \in [0,1]^d$ and $t \ge 0$ be a Kiefer process as in Definition 2. As we have seen (cf. Theorem A), this process is the limiting form of the multivariate uniform empirical process. From an earlier result of Kiefer (1961) and the strong invariance principle of Csörgő and Révész (1975), (cf. Theorem A), it follows immediately that we have

<u>Theorem H</u>: Let $K(\underline{x},t)$ be a $(d+1)$ parameter Kiefer process. Then

$$\overline{\lim_{T\to\infty}} \sup_{0\le t\le T} \sup_{\underline{x}\in[0,1]^d} \frac{|K(\underline{x},t)|}{\sqrt{\frac{1}{2}T \log\log T}} \overset{a.s.}{=} 1 .$$

For a summary and further new results of this type we refer to Révész (1976). Using the approach of Chan's paper (1976) one can also prove the following Erdős-Rényi law of large numbers (E.-R. L.L.N.) for the Kiefer process.

Theorem 1: Let a_N be a sequence of positive integers such that $a_N \to \infty$ and $a_N/N^{\delta} \to 0$ for every $\delta > 0$. Then

$$\lim_{N\to\infty} \sup_{0\le t\le N-a_N} \sup_{0\le s\le a_N} \sup_{\underline{x}\in[0,1]^d} \frac{|K(\underline{x},t+s) - K(\underline{x},t)|}{\sqrt{\frac{1}{2}a_N \log N}} \overset{a.s.}{=} 1 .$$

There is a considerable gap between Theorems H and 1. In this paper we prove a Theorem G type result for the (d+1)-parameter Kiefer process.

Let us call a monotonically non-decreasing continuous function a_T, $T \ge 0$, a β_T-function if it satisfies the conditions (i)-(iii) in Theorem G. Put $\alpha_T = (\frac{1}{2} a_T[\log \frac{T}{a_T} + \log\log T])^{-\frac{1}{2}}$. Then, if we put $a_T = T$, or if a_T is as in Theorem 1, then we can write Theorems H and 1 together as

(*) $$\overline{\lim_{T\to\infty}} \sup_{0\le t\le T-a_T} \sup_{\underline{x}\in[0,1]^d} \alpha_T|K(\underline{x},t+s) - K(\underline{x},t)| \overset{a.s.}{=} 1$$

As to the question of how near we can close the gap between Theorems H and 1, that is to say what other a_T could also satisfy (*), we state our main result here.

Theorem 2: Let a_T be a β_T-function . Then

(3.1) $$\overline{\lim_{T\to\infty}} \sup_{0\le t\le T-a_T} \sup_{\underline{x}\in[0,1]^d} \alpha_T|K(\underline{x},t+a_T) - K(\underline{x},t)| \overset{a.s.}{=} 1$$

and

(3.2) $$\overline{\lim_{T\to\infty}} \sup_{0\le t\le T-a_T} \sup_{0\le s\le a_T} \sup_{\underline{x}\in[0,1]^d} \alpha_T|K(\underline{x},t+s) - K(\underline{x},t)| \overset{a.s.}{=} 1 ,$$

where

$$(3.3) \qquad \alpha_T = (\tfrac{1}{2} a_T [\log \frac{T}{a_T} + \log\log T])^{-\frac{1}{2}} .$$

If, in addition, we also have

$$(3.4) \qquad \Sigma \exp\{-(\frac{n}{a_n})^{\varepsilon} (\frac{1}{\log n})^{1-\varepsilon}\} < \infty$$

for every $\varepsilon > 0$, then

$$(3.5) \qquad \lim_{T\to\infty} \sup_{0\le t\le T-a_T} \sup_{\underline{x}\in[0,1]^d} \alpha_T |K(\underline{x}, t+a_T) - K(\underline{x}, t)| \overset{a.s.}{=} 1$$

and

$$(3.6) \qquad \lim_{T\to\infty} \sup_{0\le t\le T-a_T} \sup_{0\le s\le a_T} \sup_{\underline{x}\in[0,1]^d} \alpha_T |K(\underline{x}, t+s) - K(\underline{x}, t)| \overset{a.s.}{=} 1 .$$

Thus, if we put $a_T = T$, we have Theorem H and, if a_N satisfies conditions in Theorem 1 then it must also satisfy (3.4) and, whence we get Theorem 1.

Another interesting case is $a_N = (\log N)^3$ and x one dimensional. In this situation (3.5) implies

$$\lim_{N\to\infty} \max_{0\le j\le N-(\log N)^3} \sup_{0\le x\le 1} \frac{|K(x, j+(\log N)^3) - K(x,j)|}{(\log N)^2} \overset{a.s.}{=} \frac{1}{\sqrt{2}} ,$$

a result which should be relevant as to the question of $r_2(n) = \log^2 n$ embedding rate of Theorem A (Komlós, Major, Tusnády 1975) being the best possible or not.

4. Some Lemmas and Proof of Theorem 2.

The proof Theorem 2 hinges on Lemma 4. First we state and prove three preliminary lemmas, leading up to the latter. The starting point is

Theorem I: (Kiefer, 1961). For each integer d and $\varepsilon > 0$ there is a constant $C(\varepsilon, d) > 0$ such that for all $\lambda \geq$ we have

$$P\{\sup_{\underline{x}\in[0,1]^d} |\alpha_n(\underline{x})| \geq \lambda\} \leq C(\varepsilon, d)\, e^{-(2-\varepsilon)\lambda^2} ,$$

where $\alpha_n(\underline{x})$ is the empirical process of Section 1.

Since $\alpha_n(\underline{x}) \xrightarrow{\mathcal{D}} B(\underline{x})$, a d-parameter Brownian bridge as in Definition 3 of Section 1, we also have

Lemma 1: Let $B(\underline{x})$, $\underline{x} \in [0,1]^d$, be a d-parameter Brownian bridge. Then, for each $\epsilon > 0$, there is a constant $C(\epsilon, d) > 0$ such that

(4.1) $$P\{\sup_{\underline{x}\in[0,1]^d} B(\underline{x}) > \lambda\} \leq C(\epsilon,d)e^{-(2-\epsilon)\lambda^2},$$

for each $\lambda > 0$.

Recalling the definition of a $(d+1)$-parameter Kiefer process $K(\underline{x},t)$, $\underline{x} \in [0,1]^d$, $t \geq 0$ from Section 1, it is seen that for each pair of positive numbers t and h, we have

$$K(\underline{x},t+h) - K(\underline{x},t) \stackrel{\mathcal{D}}{=} K(\underline{x},h) \stackrel{\mathcal{D}}{=} \sqrt{h}\, B(\underline{x})$$

for all $\underline{x} \in [0,1]^d$ and, as a consequence of Lemma 1, we get

Lemma 2: For each $\epsilon > 0$, there is a constant $C_1 = C_1(\epsilon, d)$ such that

$$P\{\sup_{\underline{x}\in[0,1]^d} (K(\underline{x},t+h) - K(\underline{x},t)) > \lambda\} \leq C_1 e^{-(2-\epsilon)\lambda^2/h},$$

for every $\lambda > 0$, $h > 0$ and $t \geq 0$. Consequently,

(4.2) $$P\{\sup_{\underline{x}\in[0,1]^d} |K(\underline{x},t+h) - K(\underline{x},t)| > \lambda\sqrt{h}\} \leq Ce^{-(2-\epsilon)\lambda^2},$$

where $C = 2C_1$.

Lemma 3: For each $\epsilon > 0$, there is a constant $C_2 = C_2(\epsilon, d) > 0$ such that

(4.3) $$P\{\sup_{0\leq s\leq h}\ \sup_{\underline{x}\in[0,1]^d} |K(\underline{x},t+s) - K(\underline{x},t)| > \lambda\sqrt{h}\} \leq C_2 e^{-(2-\epsilon)^2\lambda^2}$$

for every $h > 0$, $\lambda > 0$ and $t \geq 0$.

Proof: Since $K(\underline{x},t)$ is stationary, independent increment process in t, it suffices to show that

(4.4) $$P\{\sup_{0\leq s\leq h}\ \sup_{\underline{x}\in[0,1]^d} |K(\underline{x},s)| > \lambda\sqrt{h}\} \leq C_2 e^{-(2-\epsilon)\lambda^2}.$$

Also, $K(\underline{x},t)$ is a separable process. Hence, the countable dense set $\mathcal{D} = \bigcup \mathcal{D}_n$, where

$$\mathcal{D}_n = \{(\frac{m_1}{2^n}, \dots, \frac{m_d}{2^n}, \frac{hm_{d+1}}{2^n}) : 0 \leq m_i \leq 2^n , 1 \leq i \leq d+1\}$$

completely determines $K(\cdot,\cdot)$ on $[0,1]^d \times [0,h]$. For each fixed n, let us put $m = 1 + m_1 + m_2 2^n + \dots + m_{d+1} 2^{nd}$, for $0 \leq m_i \leq 2^n$ and $1 \leq i \leq d+1$. Define

$$K_m = K(\frac{m_1}{2^n}, \dots, \frac{m_d}{2^n}, \frac{hm_{d+1}}{2^n}) , \quad K'_m = K(\frac{m_1}{2^n}, \dots, \frac{m_d}{2^n}, h) ,$$

and

$$E_m = \{K_m > \lambda \; ; \; K'_{m'} \leq \lambda , \text{ for } m' < m\} , \; m = 1,2,\dots,2^{n(d+1)} .$$

Clearly, E_m's are disjoint and

$$\bigcup_m E_m = \{\max_{(\underline{x},t)\in\mathcal{D}_n} K(\underline{x},t) > \lambda\} .$$

Consider the event

$$B_n = \{K_m > \lambda , \; m_{d+1} = 2^n\} = \{K'_m > \lambda\} .$$

Then, because $K'_m - K_m$ is symmetric and is independent of E_m,

$$P\{E_m, B_n^C\} \leq P\{E_m, K'_m - K_m < 0\} = \tfrac{1}{2} P\{E_m\} .$$

Thus

$$P\{E_m\} \leq 2 \, P\{E_m, B_n\} . \quad \text{Since } \bigcup E_m \supseteq B_n ,$$

we have $P\{\bigcup E_m\} \leq 2P\{B_n\}$. Taking $n \to \infty$, we get, for every $\lambda > 0$.

$$P\{\sup_{0\leq s\leq h} \sup_{\underline{x}\in[0,1]^d} K(\underline{x},s) > \lambda\} \leq 2P\{\sup_{\underline{x}\in[0,1]^d} K(\underline{x},h) > \lambda\} .$$

Now, (4.4) follows from Lemma 2. This completes our proof of Lemma 3.

Now we can prove

<u>Lemma 4</u>: For each small $\varepsilon > 0$, there is a $C = C(\varepsilon, d) > 0$ such that

(4.5) $$P\{\sup_{0\le t\le T}\ \sup_{0\le s\le h}\ \sup_{\underline{x}\in[0,1]^d} |K(\underline{x}, t+s) - K(\underline{x}, t)| > \lambda h^{\frac{1}{2}}\} \le C\,\frac{T}{h}\, e^{-(2-\varepsilon)\lambda^2}$$

for every $\lambda > 0$ and $h > 0$.

<u>Proof</u>: Since $K(\underline{x}, t) \overset{\mathcal{D}}{=} \sqrt{T}\, K(\underline{x}, \frac{t}{T})$ jointly in $x \in [0,1]^d$ and $0 \le t \le T$, we need only to verify (4.5) in the case when $T = 1$. Let R be the smallest integer such that $\frac{1}{R} \le \frac{\varepsilon^2 h}{4}$ and m be the smallest integer such that $h \le \frac{m}{k} \le h + \frac{1}{k} \le h(1 + \frac{\varepsilon^2}{4}) \le h(1 + \frac{\varepsilon}{2})$. Then

$$\sup_{0\le t\le 1}\ \sup_{0\le s\le h}\ \sup_{\underline{x}\in[0,1]} |K(\underline{x}, t+s) - K(\underline{x}, t)| \le \max_{0\le n\le R}\ \sup_{1\le s\le \frac{m}{R}}\ \sup_{\underline{x}\in[0,1]^d} |K(\underline{x}, \tfrac{n}{R} + s) - K(\underline{x}, \tfrac{n}{R})| .$$

Let $\varepsilon' = \frac{\varepsilon^2}{2}$. Since $\frac{Rh}{m} > \frac{1}{1+\frac{\varepsilon}{2}}$, we have

$$\begin{aligned}
&P\{\sup_{0\le t\le 1}\ \sup_{0\le s\le h}\ \sup_{\underline{x}\in[0,1]^d} |K(\underline{x}, t+s) - K(\underline{x}, t)| > \lambda h^{\frac{1}{2}}\} \\
&\le \sum_{n=0}^{R} P\{\sup_{0\le s\le \frac{m}{R}}\ \sup_{\underline{x}\in[0,1]^d} |K(\underline{x}, \tfrac{n}{R} + s) - K(\underline{x}, \tfrac{n}{R})| > \lambda h^{\frac{1}{2}}\} \\
&\le (R+1)\, C_2(\varepsilon', d)\, \exp\{-(2-\varepsilon')\, \lambda^2 [\tfrac{Rh}{m}]\} \\
&\le (R+1)\, C_2\, \exp\{-\frac{(2-\varepsilon')}{(1+\frac{\varepsilon}{2})}\lambda^2\} \\
&\le C\, h^{-1} \exp\{-(2-\varepsilon)\lambda^2\} .
\end{aligned}$$

The last inequality holds whenever $R > 2$. This completes our proof of Lemma 4.

<u>Proof of Theorem 2</u>: Let $\varepsilon > 0$ be arbitrary and define

$$A(T) = \sup_{0\le t\le T-a_T}\ \sup_{0\le s\le T-a_T}\ \sup_{\underline{x}\in[0,1]^d} \alpha_T |K(\underline{x}, t+s) - K(\underline{x}, t)|$$

Choose $0 < \varepsilon' < 2(1 - \frac{1}{(1+\varepsilon)^2})$. Then $(2-\varepsilon')(1+\varepsilon)^2 > 2$ and let $(2-\varepsilon')(1+\varepsilon)^2 = 2 + 2\delta$, $\delta > 0$. Thus,

$$P\{A(T) > 1+\varepsilon\} \leq C(\varepsilon', d) \frac{T}{a_T} \exp\{-\log (\frac{T \log T}{a_T})^{1+\delta}\}$$

$$\leq C(\varepsilon', d) (\frac{a_T}{T})^{\delta} (\frac{1}{\log T})^{1+\delta} .$$

Put $\theta = 1 + \varepsilon$ and $T_N = \theta^N$. Then

$$\Sigma \, P\{A(T_N) > 1 + \varepsilon\} < \infty$$

and $\overline{\lim} \, A(T_N) \leq 1 + \varepsilon$ w.p.1. Observe that

$$(4.6) \qquad 1 \leq \frac{\alpha_{T_{N+1}}}{\alpha_{T_N}} \leq \theta ,$$

for large N. Since $\alpha_{T_N}^{-1} \leq \alpha_T^{-1} \leq \alpha_{T_{N+1}}^{-1}$ and

$$\alpha_{T_N}^{-1} A(T_N) \leq \alpha_T^{-1} A(T) \leq \alpha_{T_{N+1}}^{-1} A(T_{N+1}) ,$$

for $T_N \leq T \leq T_{N+1}$, it follows from (4.6) that

$$A(T) \leq \frac{\alpha_T}{\alpha_{T_{N+1}}} A(T_{N+1}) \leq (1 + \varepsilon)^2 ,$$

with probability one. This way we get

$$\overline{\lim_{T\to\infty}} \, A(T) \leq (1 + \varepsilon)^2 \quad \text{w.p.1.}$$

for every $\varepsilon > 0$. This is precisely

$$(4.7) \qquad \overline{\lim_{T\to\infty}} \, A(T) \leq 1 \quad \text{w.p.1.}$$

On the other hand, observe that

$$K(\tfrac{1}{2},1,\ldots,1,t) \overset{\mathcal{D}}{=} \tfrac{1}{2}W(t)\ ,\quad 0 \le t < \infty\ .$$

Also $\frac{1}{2}\beta_T = \alpha_T$, where β_T is defined as in Theorem G. It follows from Theorem G that

$$(4.8)\qquad \overline{\lim_{T\to\infty}} \sup_{0\le t\le T-a_T} \alpha_T|K(\tfrac{1}{2},1,\ldots,1,t+a_T) - K(\tfrac{1}{2},1,\ldots,1,t)| \ge 1 \text{ w.p.1.}\ .$$

Also

$$(4.9)\qquad \underline{\lim_{T\to\infty}} \sup_{0\le t\le T-a_T} \alpha_T|K(\tfrac{1}{2},1,\ldots,1,t+a_T) - K(\tfrac{1}{2},1,\ldots,1,t)| \ge 1 \text{ w.p.1.}\ ,$$

whenever (3.4) is satisfied. The statement of Theorem 2 now follows from (4.7)-(4.9).

One can similarly prove the following P. Lévy modulus of continuity for a Kiefer process.

Theorem 3: Let $K(\cdot,\cdot)$ be a Kiefer process as in Theorem 2. Then

$$\lim_{h\to 0} \sup_{0\le t\le 1-h} \sup_{x\in[0,1]^d} \frac{K(x,t+h) - K(x,t)}{\sqrt{h\log h^{-1}}} \overset{\text{a.s.}}{=} \frac{1}{\sqrt{2}}\ .$$

REFERENCES

[1] Chan, Arthur H.C. (1976). Erdős-Rényi type modulus of continuity theorems for Brownian sheets. To appear.

[2] Chan, A.H.C., Csörgő, M. and Révész, P. (1976). On the increments of a multi-parameter Wiener process. To appear.

[3] Csörgő, M. and Révész, P. (1975). A new method to prove Strassen type laws of invariance principle II. Z. Wahrscheinlichkeitstheorie verw. Geb. 31 261-269.

[4] Csörgő, M. and Révész, P. (1976a). A strong approximation of the multivariate empirical process. To appear.

[5] Csörgő, M. and Révész, P. (1976b). How big are the increments of a Wiener process? To appear.

[6] Erdős, P. and Rényi, A. (1970). On a new law of large numbers. Journal d'Analyse Mathematique 13 103-111.

[7] Kiefer, J. (1961). On large deviations of the empiric d.f. of vector chance variables and a law of the iterated logarithm. Pacific J. Math. 11 649-660.

[8] Komlós, J., Major, P. and Tusnády, G. (1975). An approximation of partial sums of independent RV's, and the sample DF.I. Z. Wahrscheinlichkeitstheorie verw. Geb. 32 111-131.

[9] Komlós, J., Major, P. and Tusnády, G. (1975). An approximation of partial sums of independent RV's, and the sample DF. II. Z. Wahrscheinlichkeitstheorie verw. Geb. 34 33-58.

[10] Lévy, P. (1937). Theorie de l'addition des variables aleatoires independantes, Aguthier-Villars, Paris.

[11] Major, P. (1976). The approximation of partial sums of independent r.v.'s. To appear.

[12] Müller, D.W. (1970). On Glivenko-Cantelli convergence. Z. Wahrscheinlichkeitstheorie verw. Geb. 16 195-210.

[13] Orey, S. and Pruitt, W.E. (1973). Sample functions of the N-parameter Wiener process. Ann. Probability 1 138-163.

[14] Orey, S. and Taylor, S.J. (1974). How often on a Brownian paths does the law of iterated logarithm fail? Proc. London Math. Soc. (3) 28 174-192.

[15] Révész, P. (1976). Three theorems on the multivariate empirical process. In these Lecture Notes.

[16] Kiefer, J. (1972). Skorohod embedding of multivariate RV's and the sample DF. Z. Wahrscheinlichkietstheorie verw. Geb. 24 1-35.

Kolmogorov-Smirnov tests when parameters are estimated

J. Durbin
London School of Economics and Political Science

1. Introduction and summary

In this paper we review a number of techniques for constructing tests of goodness of fit of Kolmogorov-Smirnov type when parameters are estimated from the data. Our main concern will be with asymptotic theory.

The limiting distribution of the sample distribution function when parameters are estimated is considered in section 2. Since this differs from the corresponding distribution when the parameters are known, the Kolmogorov-Smirnov tests appropriate to the parameter-known case are invalid, even as approximations, for the unknown-parameter case.

In section 3 the method of random substitution suggested by Durbin (1961) will be considered from an asymptotic point of view. It is shown that when the method is used the limiting distribution of the sample process on the null hypothesis is the same as if the values of the nuisance parameters were known. The limiting distribution under a specified sequence of alternative hypotheses is obtained.

Section 4 examines the limiting performance of the half-sample device suggested by Durbin (1973b). In this device the unknown parameter vector is estimated from a randomly-chosen half-sample of data and the sample distribution function (df) is constructed as if the estimate were the true value. It is shown that the limiting distribution of the sample df so obtained is the same as that given by the random substitution method under both the null hypothesis and under the sequence of alternative hypotheses. The two methods are therefore asymptotically equivalent. The half-sample device is, however, easier to apply in practice.

Section 4 refers briefly to some Monte-Carlo studies of the distributions of Kolmogorov-Smirnov statistics for testing normality and exponentiality due to Lilliefors (1967, 1969) and Stephens (Table 54 of Pearson and Hartley (1972)) while section 5 mentions some exact results of Durbin (1975).

In section 6 an analogue of the reflection method appropriate to the parameter-unknown case is developed. This requires the use of a random boundary instead of the fixed boundary used for the usual Kolmogorov-Smirnov tests. Because of this, the practical value of the method is questionable owing to the amoung of computing required to implement it.

2. Limiting distribution of the sample process when parameters are estimated

Suppose that $x_1,\ldots,x_n$ are independent observations from a distribution with continuous df $F(x,\theta)$ where θ is a vector of parameters. Assume first that θ is known to equal θ_0. Let $F_n(t)$ be the proportion of the values $x_1,\ldots,x_n$ such that $F(x_j,\theta_0) \leq t$. It is well known that the sample process defined by $y_n(t) = \sqrt{n}\,[F_n(t) - t]$ for $0 \leq t \leq 1$ converges weakly to the tied down Brownian motion process, i.e. the normal process $y(t)$ with mean zero and covariance function

$$E[y(t_1)y(t_2)] = \min(t_1,t_2) - t_1t_2\,, \quad 0 \leq t_1,\, t_2 \leq 1. \tag{2.1}$$

Many exact and asymptotic results are known about the distributions of the Kolmogorov-Smirnov statistics

$$D_n^+ = \sup_{0\leq t\leq 1}[F_n(t)-t]\,, \quad D_n^- = \sup_{0\leq t\leq 1}[t-F_n(t)]\,, \quad D_n = \max(D_n^+,D_n^-) \tag{2.2}$$

for this case.

In this paper we are interested in the extension to the case $\theta = [\theta_1',\ \theta_2']'$ where θ_1 is a vector of p_1 parameters specified on the null hypothesis as equal to θ_{10} while θ_2 is a vector of nuisance parameters whose true value θ_{20} is unknown. Let $\hat{\theta}_{2n}$ be an estimator of θ_{20}, let $\hat{\theta}_n = [\theta_{10}',\ \hat{\theta}_{2n}']'$, let $\hat{F}_n(t)$ be the proportion of $x_1,\ldots,x_n$ such that $F(x_j,\ \hat{\theta}_n) \leq t$ and let $\hat{y}_n(t) = \sqrt{n}\,[\hat{F}_n(t)-t]$ for $0 \leq t \leq 1$. We wish to consider the limiting distribution of $\hat{y}_n(t)$ under the sequence of alternatives $H_n : \theta_{1n} = \theta_{10} + n^{-\frac{1}{2}}\gamma$ where γ is a given vector.

We assume that the density $f(x,\theta) = \partial F(x,\theta)/\partial x$ exists for almost all x and for all θ in a closed neighbourhood of $\theta_0 = [\theta_{10}',\theta_{20}']'$ and that $\hat{\theta}_{2n}$ is a regular efficient estimator such that

$$\sqrt{n}\,(\hat{\theta}_{2n}-\theta_{20}) = \frac{1}{\sqrt{n}}\,\mathcal{I}^{-1}\sum_{j=1}^{n}\frac{\partial \log f(x_j,\theta_n)}{\partial\theta_2} + \mathcal{I}^{-1}\,\mathcal{I}_{21}\,\gamma + \varepsilon_{1n} \tag{2.3}$$

where $\partial \log f(x_j,\theta_n)/\partial\theta_2$ is the vector of derivatives of $\log f(x_j,\theta)$ with respect to the elements of θ_2 evaluated at $\theta_n = [\theta_{1n}', \theta_{20}']'$,

$$\mathcal{I} = E\left[\frac{\partial \log f(x,\theta_0)}{\partial\theta_2} \frac{\partial \log f(x,\theta_0)}{\partial\theta_2'} \middle| \theta = \theta_o\right],$$

$$\mathcal{I}_{21} = E\left[\frac{\partial \log f(x,\theta_0)}{\partial\theta_2} \frac{\partial \log f(x,\theta_0)}{\partial\theta_1'} \middle| \theta = \theta_0\right]$$

the derivatives being evaluated at $\theta = \theta_0$, and where $\varepsilon_{1n} \overset{p}{\to} 0$.

Under specified regularity conditions on $F(x,\theta)$, Durbin (1973a) showed that $\hat{y}_n(t)$ converges weakly to a process $\hat{y}(t)$ which is normal with mean function

$$E[\hat{y}(t)] = \gamma'[g_1(t) - \mathcal{I}_{21}' \mathcal{I}^{-1} g_2(t_2)] \tag{2.4}$$

and covariance function

$$C[\hat{y}(t_1),\hat{y}(t_2)] = \min(t_1,t_2) - t_1t_2 - g_2(t_1)'\mathcal{I}^{-1} g_2(t_2) \tag{2.5}$$

where $g_i(t) = \partial F(x,\theta)/\partial\theta_i$ when this is expressed as a function of $t = F(x,\theta_0)$ for $i = 1,2$, the derivatives being evaluated at $\theta = \theta_0$.

Since the covariance functions (2.1) and (2.5) are different the processes $y_n(t)$ and $\hat{y}_n(t)$ have different limiting distributions. The effect is that the asymptotic distributions of the Kolmogorov-Smirnov statistics (2.2) differ according as θ_2 is or is not estimated from the data. In consequence results already available for the process $y(t)$ cannot be used to provide results for the process $\hat{y}(t)$, even as approximations. It turns out that the problem of finding limiting distributions of Kolmogorov-Smirnov statistics for the process $\hat{y}(t)$ is extremely difficult. In the next two sections we therefore follow the alternative strategy of transforming the problem into a form for which the results for process $y(t)$ are applicable.

3. The random substitution method

The object of this technique, proposed for the finite-sample case by Durbin (1961), is to transform tests of composite hypotheses into tests of simple hypotheses. This is done by replacing the efficient estimator $\hat{\theta}_{2n}$ of the unknown θ_{20} calculated from the sample by a corresponding "estimator", external to the sample, of a known value of θ_2.

Taking the finite-sample case first, suppose that on H_o a sufficient statistic T_1 for θ_2 exists and that a further statistic T_2 exists which is distributed independently of T_1 such that the transformation $\tau : x_1,\dots,x_n \to T_1,T_2$ has a unique inverse τ^{-1}, i.e. given the values of T_1 and T_2 the sample values $x_1,\dots,x_n$ can be constructed from them uniquely. Since T_1 is sufficient the distribution of T_2 does not depend on θ_2. Suppose that on H_o T_1 is known to have distribution function $G(T_1,\theta_2)$. Let θ_2^* be an arbitrarily selected value of θ_2 and let T_1^* be a random vector from the distribution with distribution function $G(T_1^*,\theta_2^*)$. Let $x_1^*,\dots,x_n^*$ be the values of $x_1,\dots,x_n$ determined from T_1^*, T_2 by applying the inverse transformation τ^{-1}. Then on H_o $x_1^*,\dots,x_n^*$ are independent random variables from the distribution with known distribution function $F(x,\theta^*)$ where $\theta^* = [\theta_{10}',\theta_2^{*\prime}]'$. The composite hypothesis $H_o : \theta_1 = \theta_{10}$, θ_2 unknown, based on $x_1,\dots,x_n$ may therefore be replaced by the simple hypothesis $H_o^* : \theta_1 = \theta_{10}$, $\theta_2 = \theta_2^*$, based on $x_1^*,\dots,x_n^*$.

In this paper we consider from a heuristic point of view an asymptotic form of this method. Since $\hat{\theta}_{2n}$ as defined in the last section attains the minimum-variance bound asymptotically let us take $T_1 = \hat{\theta}_{2n}$. On H_n, let $F_n(t)$ be the proportion of $x_1,\dots,x_n$ such that $F(x_j,\theta_n) \leq t$ and let $y_n(t) = \sqrt{n}[F_n(t)-t]$ for $0 \leq t \leq 1$. To find the appropriate T_2 we note that on H_n Durbin (1973a) Lemma 2 showed effectively that

$$\hat{y}_n(t) = y_n(t) + \gamma' g_1(t) - \sqrt{n}(\hat{\theta}_{2n}-\theta_{20})g_2(t) + \varepsilon_{2n}(t)$$

so that on H_o,

$$\hat{y}_n(t) = y_n(t) - \sqrt{n}(\hat{\theta}_{2n}-\theta_{20})\, g_2(t) + \varepsilon_{2n}(t) \qquad (3.1)$$

where $\varepsilon_{2n}(t) \xrightarrow{p} 0$ uniformly. A straightforward calculation shows that $\hat{\theta}_{2n}$ and $\hat{y}_n(t)$ are asymptotically uncorrelated. They are also asymptotically jointly normal and hence independent. We therefore take $T_2 = \hat{y}_n(t)$, $0 \leq t \leq 1$.

The observations $x_1,\dots,x_n$ are mathematically equivalent to the sample process $y_n(t)$ for $0 \leq t \leq 1$. For sample size n denote the transformation τ from $y_n(t)$ to $\hat{\theta}_{2n}$, $\hat{y}_n(t)$ by τ_n. The inverse transformation is τ_n^{-1}: $\hat{\theta}_{2n}$, $\hat{y}_n(t) \to y_n(t)$. From (3.1) this has the form

$$y_n(t) = \hat{y}_n(t) + \sqrt{n}(\hat{\theta}_{2n}-\theta_{20})g_2(t) - \varepsilon_{2n}(t) \qquad (3.2)$$

Suppose that on H_o $\hat{\theta}_{2n}$ has distribution function $G_n(\hat{\theta}_{2n},\theta_{20})$.

Let θ^*_{2n} be a random vector independent of $x_1,\ldots,x_n$ from a distribution with distribution function $G_n(\theta^*_{2n},\theta^*_2)$ where θ^*_2 is an arbitrarily selected value of θ_2. By analogy with (3.2) define

$$y^*_n(t) = \hat{y}_n(t) + \sqrt{n}(\theta^*_{2n}-\theta^*_2)g^*_2(t) - \varepsilon^*_{2n}(t) \tag{3.3}$$

where $g^*_2(t) = \partial F(x,\theta)/\partial\theta_2$, the vector of derivatives being evaluated at $\theta = \theta^* = [\theta'_{10}, \theta^{*\prime}_2]'$, and where $\varepsilon^*_{2n}(t) \xrightarrow{p} 0$ uniformly. Alternatively, suppose that $y^*_n(t)$ is the sample process computed from $x^*_1,\ldots,x^*_n$ obtained as described at the beginning of this section and assume that (3.3) holds.

Inspection of (2.5) shows that the part of the covariance function involving θ_2 is $g_2(t_1)'\,\mathcal{J}^{-1}\,g_2(t_2)$. From now on let us assume that this quantity is independent of θ_2. At first sight this may appear to be a rather strong assumption. However, it seems to hold under rather general conditions although I have not succeeded in formulating these conditions in a satisfactory form. As an example, the assumption holds when the density of x on H_o has the form

$$f(x,\theta_2) = \frac{1}{\theta_{22}}\, g\left(\frac{x-\theta_{21}}{\theta_{22}}\right)$$

where $\theta_2 = [\theta_{21},\theta_{22}]'$, as is easily verified. This simple example includes a number of situations important in practice.

Now

$$\lim_n n\, E[(\theta^*_{2n}-\theta^*_2)(\theta^*_{2n}-\theta^*_2)'] = \mathcal{J}^{*-1}$$

where

$$\mathcal{J}^* = E\left[\frac{\partial \log f(x,\theta^*)}{\partial\theta^*_2}\;\frac{\partial \log f(x,\theta^*)}{\partial\theta^{*\prime}_2}\right].$$

Consequently,

$$\lim_n nE[g^*_2(t_1)'(\theta^*_{2n}-\theta^*_2)(\theta^*_{2n}-\theta^*_2)'g^*_2(t_2)] = g^*_2(t_1)'\mathcal{J}^{*-1}\,g^*_2(t_2)$$

which $= g_2(t_1)\,\mathcal{J}^{-1}\,g_2(t_2)$ since this has been assumed to be independent of θ_2. Furthermore $\lim_n E[\sqrt{n}(\theta^*_{2n}-\theta^*_2)] = 0$. Since θ^*_{2n} is independent of $\hat{y}_n(t)$ and $\hat{\theta}_{2n}$ it follows from (3.3) that on H_n,

$$\lim_n E[y^*_n(t)] = \gamma'[g_1(t) - \mathcal{J}'_{21}\,\mathcal{J}^{-1}\,g_2(t)] \tag{3.4}$$

$$\lim_n C[y_n^*(t_1), y_n^*(t_2)] = \min(t_1, t_2) - t_1 t_2 . \tag{3.5}$$

We deduce that on H_o, $y_n^*(t)$ has the same limiting distribution as $y_n(t)$. Consequently, test statistics such as D_n^+, D_n^- and D_n have the same limiting distributions when calculated from $y_n^*(t)$ as when they have been calculated from $y_n(t)$ for the purpose of testing simple hypotheses. For example, if D_n^* is the value of D_n calculated from $y_n^*(t)$ then

$$\lim_n \Pr[\sqrt{n}\, D_n^* \leq d] = \sum_{j=-\infty}^{\infty} (-1)^j e^{-2j^2 d^2} .$$

At first sight the random-substitution technique seems to offer a relatively simple solution to the problem under discussion but it has not found favour in practice. The reasons seem to be

(1) Practical workers do not like introducing into the analysis of real data the element of artificial randomisation involved in selecting the random vector θ_{2n}^* independently of the data.

(2) It is suspected that the use of randomisation must inevitably entail a loss of power relative to the corresponding test based on $\hat{y}_n(t)$.

(3) The computational labour required to obtain θ_{2n}^* and effect the required transformations is burdensome.

As regards (2) it is interesting to note that in the important case in which $\mathfrak{I}_{21} = 0$ the limiting distribution of $y_n^*(t)$ is the same as that of $y_n(t)$ under the sequence of alternatives H_n as well as under H_o. This implies that the limiting power of the test based on $y_n^*(t)$ is the same as that obtained from $y_n(t)$ when the true value of the nuisance parameter vector θ_2 is known. Of course, this is not to say that greater limiting power cannot be obtained, whether θ_2 is known or not, by estimating it by $\hat{\theta}_{2n}$ and basing the test on $\hat{y}_n(t)$. The case $\mathfrak{I}_{21} = 0$ is that in which maximum-likelihood estimates of θ_1 and θ_2 are asymptotically uncorrelated.

So far as (3) is concerned we shall consider in the next section a much simpler way of constructing a form of the sample process which has the same limiting distribution as $y_n^*(t)$ under both null and alternative hypotheses.

4. The half-sample device

The purpose of the half-sample device is the same as that of the asymptotic form of the random substitution method, namely to obtain a version of the sample process whose limiting distribution in the composite-hypothesis case is the same as that of the usual form of the sample process in the simple-hypothesis case. Remarkably, it turns out that under general conditions this objective can be achieved simply by calculating the estimate of θ_2 from a randomly-chosen half-sample of the data instead of from the whole sample. The device was suggested by Durbin (1973b) following up an earlier related proposal by Rao (1972).

Suppose that $n = 2m$ and that the observations $x_1,\ldots,x_n$ are in random order. Let $\tilde{\theta}_{2n}$ be an efficient regular estimator of θ_{20} derived from the half-sample $x_1,\ldots,x_m$, i.e.

$$\sqrt{m}(\tilde{\theta}_{2n}-\theta_{20}) = \frac{1}{\sqrt{m}}\mathcal{I}^{-1}\sum_{j=1}^{m}\frac{\partial \log f(x_j,\theta_n)}{\partial\theta_2} + \mathcal{I}^{-1}\mathcal{I}_{21}\gamma + \varepsilon_{3n}$$

where $\varepsilon_{3n} \xrightarrow{p} 0$. Let $\tilde{F}_n(t)$ be the proportion of $x_1,\ldots,x_n$ such that $F(x_j,\tilde{\theta}_{2n}) \leq t$ and let $\tilde{y}_n(t) = \sqrt{n}[\tilde{F}_n(t)-t]$, $0 \leq t \leq 1$. We emphasise that $\tilde{F}_n(t)$ is calculated from the full set of data $x_1,\ldots,x_n$ and not just from $x_1,\ldots,x_m$. Let

$$\tilde{w}_n = \frac{2}{\sqrt{n}}\mathcal{I}^{-1}\sum_{j=1}^{m}\frac{\partial \log f(x_j,\theta_n)}{\partial\theta_2} .$$

Lemmas 1 and 2 of Durbin (1973a) imply that on H_n,

$$\tilde{y}_n(t) = y_n(t) + \gamma'[g_1(t) - \mathcal{I}_{21}'\mathcal{I}^{-1}g_2(t)] - \tilde{w}_n' g_2(t) + \varepsilon_{4n}(t) \tag{4.1}$$

where $\varepsilon_{4n}(t) \xrightarrow{p} 0$ uniformly. Since $E[y_n(t)] = 0$ and $E[\tilde{w}_n] = 0$ we have

$$\lim_n E[\tilde{y}_n(t)] = \gamma'[g_1(t) - \mathcal{I}_{21}'\mathcal{I}^{-1} g_2(t)] . \tag{4.2}$$

Let $d_j = 1-t$ if $F(x_j,\theta_n) \leq t$ and $d_j = -t$ otherwise and let $y_n^{(1)}(t) = n^{-1/2}\sum_{j=1}^{m} d_j$, $y_n^{(2)}(t) = n^{-1/2}\sum_{j=m+1}^{n} d_j$. Then $y_n(t) = y_n^{(1)}(t) + y_n^{(2)}(t)$. As in the proof of Lemma 2 of Durbin (1973a), $E[y_n^{(1)}(t)\tilde{w}_n] = \mathcal{I}^{-1}g_2(t)$. Also, $E[y_n^{(2)}\tilde{w}_n] = 0$ and $E[\tilde{w}_n\tilde{w}_n'] = 2\mathcal{I}^{-1}$. Consequently, from (4.1) we obtain

$$\lim_n C[\tilde{y}_n(t_1), \tilde{y}_n(t_2)] = \min(t_1,t_2) - t_1t_2 . \tag{4.3}$$

The limiting normality of $\tilde{y}_n(t)$ follows from the treatment of Durbin (1973a).

We note that (4.2) and (4.3) are the same as (3.4) and (3.5). Thus the limiting distributions of $y_n^*(t)$ and $\tilde{y}_n(t)$ are the same on both the null and alternative hypotheses. It follows that the random substitution and half-sample methods are asymptotically equivalent. The half-sample device is intuitively more appealing than the random substitution method since it does not entail going outside the set of observed values $x_1,\ldots,x_n$. Moreover, it is computationally easier since a computing procedure has to be available for the calculation of $\hat{\theta}_{2n}$ in any case and this can easily be applied to give $\tilde{\theta}_{2n}$. On the other hand, the randomisation element required to select the half-sample is necessarily equivalent to that involved in the use of the random substitution method. Moreover, in finite samples the half-sample device is at best approximate whereas the random substitution method can give an exact test.

The asymptotic equivalence of the random-substitution method and the half-sample device was established for a different class of tests in Durbin (1975a).

Should the half-sample technique be recommended for practical use? Obviously, if the corresponding full-sample procedure is available and convenient this should be used in preference. If, however, appropriate full-sample procedures are either unavailable or inconvenient the half-sample device might provide a useful interim procedure.

5. Monte-Carlo solutions

In some important practical cases the finite-sample distribution of $\hat{F}_n(t)$ is independent of unknown parameters. Consequently it is possible to simulate the distribution of appropriate test statistics such as the Kolmogorov-Smirnov statistics by Monte-Carlo methods and to prepare tables of significance points from the observed distributions. This was done first for tests of normality and tests of exponentiality by Lilliefors (1967,1969) by means of Monte-Carlo experiments of rather modest size. Much more extensive experiments were carried out for both normal and exponential cases by M. A. Stephens. Stephens' results, enhanced by sophisticated smoothing and other devices, were embodied in extremely compact form in Table 54 of Pearson and Hartley (1972).

6. Exact results

In favourable cases it is possible to make some progress with the finite-sample distributions of the Kolmogorov-Smirnov statistics. Let $\hat{t}_j = F(x_j,\hat{\theta}_n)$ for $j = 1,\ldots,n$. For a wide class of tests of Kolmogorov-Smirnov type the acceptance probability takes the form

$$\hat{P} = \Pr[u_j \leq \hat{t}_j \leq v_j, \quad j = 1,\dots,n]$$

for suitable $u_1, v_1, \dots u_n, v_n$. In a number of important cases the distribution of $\hat{t}_1, \dots, \hat{t}_n$ is independent of both θ_{20} and $\hat{\theta}_{2n}$. For this situation Durbin (1975b) showed that, when the densities exist,

$$P = \frac{g_r(\theta_2^*)}{g_u(\theta_2^*)}$$

where $g_r(\theta_2^*)$ is the density of $\hat{\theta}_{2n}$ at an arbitrary value θ_2^* arising from samples restricted to satisfy $u_j \leq t_j \leq v_j$ where $t_j = F(x_j, \theta^*)$, $\theta^* = [\theta_{10}, \theta_2^{*\prime}]'$, $j = 1,\dots,n$, and where $g_u(\theta_2^*)$ is the unrestricted density of $\hat{\theta}_{2n}$ at θ_2^*. For estimators of the form $\hat{\theta}_{2n} = n^{-1} \sum_{j=1}^{n} h_j(x_j)$, techniques were developed for the computation and and inversion of the Fourier transform of $g_r(\theta_2^*)$. These were applied to the construction of tables of percentage points for D_n^+, D_n^- and D_n for the test of the hypothesis that the data come from an exponential distribution. The exponential distribution is, of course, a particularly simple case and it must be acknowledged that to extend the application of these techniques to other distributions would be very onerous.

7. A modified reflection method

The easiest and most transparent way to obtain the limiting distributions of the standard Kolmogorov-Smirnov statistics for the simple-hypothesis case is to use a form of the reflection method, e.g. as in Doob (1949) or Durbin (1973b), chapter 4. It is natural therefore to consider whether the technique can be employed in the estimated-parameter case. Now the applicability of the reflection method depends essentially on the Markovian nature of the process under study and a brief examination of the properties of the process $\hat{y}(t)$ suffices to reveal that unlike $y(t)$ it is not a Markov process. It is therefore clear that the reflection method cannot be applied to $\hat{y}(t)$ directly. Instead we have to modify the problem in such a way that any Markovian properties it possesses can be exploited in a suitable fashion. One way of doing so is the following.

Let $w(t)$, $t \geq 0$, be the Brownian motion process such that $E[w(t)] = 0$, $E[w(t_1)w(t_2)] = \min(t_1,t_2)$, let $\dot{g}_2(t) = dg_2(t)/dt$ and let $b(t) = \int_0^t \dot{g}_2(s)\, dw(s)$ for $0 \leq t \leq 1$, the integral being interpreted as mean-square limit. Let $z(t) = w(t) - tw(1) - g_2(t)\, \mathcal{I}^{-1}\, b(1)$. It is easy to verify both that $z(t)$ is distributed as $w(t)$ given $w(1)=b(1)=0$

and also that $z(t)$ has mean zero and covariance function (2.5). It follows that $\hat{y}(t)$ is distributed as $z(t)$ and hence as $w(t)$ given $w(1)= b(1) = 0$.

We now exploit the fact that although $z(t)$ itself is not a Markov process, $w(t)$ and $b(t)$ are jointly Markovian. For the construction of a one-sided test analogous to that based on D_n^+ our approach is to seek a random boundary $y = a(t)$, where $a(t)$ is a function of $z(s)$ for $s \leq t$, such that the probability that a sample path of $z(t)$ crosses this boundary is e^{-2d^2} where d is a given constant. This is the same thing as seeking $a(t)$ depending on $w(s)$ for $s \leq t$ such that the probability that a sample path of $w(t)$ crosses $y = a(t)$ given $w(1) = b(1) = 0$ is e^{-2d^2}.

Let $t = r$ be the first time, if any, that a particular sample path of $w(t)$ meets the boundary $y = a(t)$. Our proposal is that $a(r)$ should be chosen so that given the previous history of $w(t)$ and $b(t)$ for $t \leq r$, the conditional density of $w(1)$ given $b(1) = 0$ at $w(1) = 2d$ should be equal the conditional density of $w(1)$ given $b(1) = 0$ at $w(1) = 0$. Since $w(t)$ is a strong Markov process it follows that $a(r)$ is a function of $b(r)$ only. Our objective will be achieved if the conditional mean of $w(1)$ given $b(r)$, given $w(r) = a(r)$ and given $b(1) = 0$ is equal to d.

Now the regression coefficient vector of $w(1) - w(r)$ on $b(1) - b(r)$ is

$$\delta = \left[\int_r^1 \dot{g}_2(t)\dot{g}_2(t)'dt\right]^{-1} \int_r^1 \dot{g}_2(t)dt$$

$$= - [v(1) - v(r)]^{-1} g_2(r)$$

where $v(s) = \int_0^s \dot{g}_2(t)\dot{g}_2(t)'dt$ for $0 \leq s \leq 1$ since $g_2(1) = 0$. When $b(1) = 0$ and $b(r)$ is given, $b(1)-b(r) = -b(r)$. The conditional mean of $w(1)-w(r)$ is therefore $g_2(r)'[v(1)-v(r)]^{-1}b(r)$. This has to equal $d-a(r)$. We therefore have

$$a(r) = d - g_2(r)'[v(1)-v(r)]^{-1} b(r), \quad 0 < r < 1 . \tag{7.1}$$

To apply this result to $\hat{y}(t)$, $b(r)$ is replaced by $\hat{b}(r) = \int_0^r \dot{g}_2(t)d\hat{y}(t)$. The condition $\hat{b}(1) = 0$ is, of course, satisfied automatically.

To prove that this construction gives a rejection probability of e^{-2d^2} we note that the probability that a sample path of $w(t)$ crosses the boundary for the first time at $t = r$ and $0 \leq w(1) \leq \varepsilon$ given $b(1)=0$ is the same as the probability that a sample path crosses the boundary for the first time at $t = r$ and $2d - \varepsilon \leq w(1) \leq 2d$ given $b(1) = 0$. Integrating with respect to r, the probability that a sample path crosses the boundary and $0 \leq w(1) \leq \varepsilon$ given $b(1) = 0$ is the same as the probability that it crosses the boundary and $2d - \varepsilon \leq w(1) \leq 2d$ given $b(1) = 0$. But for small enough ε, every path satisfying $2d - \varepsilon \leq w(d) \leq 2d$ crosses the boundary. Consequently the probability that a sample path crosses the boundary and $0 \leq w(1) \leq \varepsilon$ given $b(1) = 0$ equals the probability that $2d - \varepsilon \leq w(1) \leq 2d$ given $b(1) = 0$ and this equals $(2\pi)^{-\frac{1}{2}} e^{-\frac{1}{2}(2d)^2} + o(\varepsilon) = (2\pi)^{-\frac{1}{2}} e^{-2d^2} + o(\varepsilon)$ since $w(1)$ is $N(0,1)$ and $w(1)$ is independent of $b(1)$. The probability that $w(t)$ crosses the boundary given $w(1) = 0$ and $b(1) = 0$ is therefore

$$\lim_{\varepsilon\to 0} \frac{\Pr[\text{a path crosses the boundary and } 0 \leq w(1) \leq \varepsilon \mid b(1) = 0]}{\Pr[0 \leq w(1) \leq \varepsilon \mid b(1) = 0]}$$

$$= \lim_{\varepsilon\to 0} \frac{(2\pi)^{-\frac{1}{2}} e^{-2d^2} + o(\varepsilon)}{(2\pi)^{-\frac{1}{2}} + o(\varepsilon)}$$

$$= e^{-2d^2} .$$

For computational purposes with a finite sample we note that

$$g_2(t) = \frac{d\, g_2(t)}{dt} = \frac{d\, g_2(t)}{dx}\frac{dx}{dt} = \frac{d}{dx}\frac{\partial F(x,\theta)}{\partial\theta_2}\frac{1}{f(x,\theta)} = \frac{\partial \log f(x,\theta)}{\partial\theta_2} .$$

Consequently, taking $\hat{y}_n(r)$ in place of $\hat{y}(r)$ we obtain

$$\hat{b}(r) = \int_0^r \dot{g}_2(t)\, d\hat{y}_n(r) = \sqrt{n} \int_0^r \dot{g}_2(t)\, [d\hat{F}_n(t) - dt]$$

$$= \frac{1}{\sqrt{n}} \sum_{x_j \leq x} \frac{\partial \log f(x_j, \hat{\theta}_n)}{\partial\theta_2} - \sqrt{n}\, \frac{\partial F(x, \hat{\theta}_n)}{\partial\theta}$$

where $r = F(x, \hat{\theta}_n)$. We see that by pursuing this line of work a computationally feasible procedure could be worked out. However, it is not suggested that this possibility should be considered seriously from the standpoint of practical work in statistics in view of the labour involved in constructing the boundary in any particular case.

An essentially identical construction could be used for a test based on D_n^-.

References

Durbin, J. (1961). Some methods of constructing exact tests. *Biometrika*, **48**, 41-55.

Durbin, J. (1973a). Weak convergence of the sample distribution function when parameters are estimated. *Ann. Statist.*, **1**, 279-290.

Durbin, J. (1973b). *Distribution Theory for Tests based on the Sample Distribution Function*. Philadelphia: Society of Industrial and Applied Mathematics.

Durbin, J. (1975a). Tests of model specification based on residuals. *A Survey of Statistical Design and Linear Models* ed. J. N. Srivastava. Rotterdam: North Holland.

Durbin, J. (1965b). Kolmogorov-Smirnov tests when parameters are estimated with applications to tests of exponentiality and tests on spacings. *Biometrika*, **62**, 5-22.

Lilliefors, H.W. (1967). On the Kolmogorov-Smirnov test for normality with mean and variance unknown. *J.Am.Statist.Assoc.*, **62**, 399-402.

Lilliefors, H.W. (1969). On the Kolmogorov-Smirnov test for the exponential distribution with mean unknown. *J.Am.Statist.Assoc.*, **64**, 387-389.

Pearson, E.S. and Hartley, H.O. (1972). *Biometrika Tables for Statisticians*, Vol. 2. Cambridge University Press.

Rao, K.C. (1972). The Kolmogoroff, Cramér-von Mises, Chisquare statistics for goodness-of-fit tests in the parametric case. (Abstract). *Bull.Inst.Math.Statist.*, **1**, 87.

ON UNIFORM CONVERGENCE OF MEASURES WITH APPLICATIONS TO UNIFORM CONVERGENCE OF EMPIRICAL DISTRIBUTIONS

P. Gaenssler and W. Stute
Ruhr University, Math. Inst. NA
D-4630 Bochum, West Germany

1. Introduction. Throughout this paper let $(\xi_i)_{i\in\mathbb{N}}$ be a sequence of independent identically distributed (i.i.d.) random vectors defined on some probability space (p-space) $(\Omega, \mathcal{F}, \mathbb{P})$ and taking their values in the k-dimensional ($k\geq 1$) Euclidean space $\mathbb{R}^k$. Let $\mu = Q_{\xi_1}$ and $F = F_{\xi_1}$, respectively, denote the probability distribution (p.d.) of ξ_1 on the Borel σ-field $\mathcal{B}_k$ in $\mathbb{R}^k$ and distribution function (d.f.) of ξ_1 on $\mathbb{R}^k$. Then, given n observations $\xi_1(\omega),\ldots,\xi_n(\omega)$, $n\in\mathbb{N}$, $\omega\in\Omega$, let $\mu_n = \mu_n^\omega$ and $F_n = F_n^\omega$ be the corresponding empirical p.d. on $\mathcal{B}_k$ and d.f. on $\mathbb{R}^k$, respectively, i.e.

(1.1) $\mu_n^\omega(C) := \frac{1}{n}\sum_{i=1}^{n}\varepsilon_{\xi_i(\omega)}(C) = \frac{1}{n}|\{1\leq i\leq n: \xi_i(\omega)\in C\}|$, $C\in\mathcal{B}_k$, and

(1.2) $F_n^\omega(\underline{t}) := \mu_n^\omega((-\underline{\infty},\underline{t}\,])$, $\underline{t}\in\mathbb{R}^k$,

where ε_x denotes the Dirac measure in $x\in\mathbb{R}^k$ and $(-\underline{\infty},\underline{t}\,] := \prod_{i=1}^{k}(-\infty,t_i]$ for $\underline{t}=(t_1,\ldots,t_k)\in\mathbb{R}^k$. For arbitrary but fixed $C\in\mathcal{B}_k$ the random variables $\eta_i := \chi_C\circ\xi_i$ (where χ_C denotes the indicator function of C), $i\in\mathbb{N}$, are again i.i.d., hence by the strong law of large numbers

(1.3) $\mu_n^\omega(C) = \frac{1}{n}\sum_{i=1}^{n}\eta_i(\omega) \xrightarrow[n\to\infty]{} E(\eta_1) = \mu(C)$ for $\mathbb{P}$-a.a. $\omega\in\Omega$,

i.e. neglecting a $\mathbb{P}$-null set (which may depend on C) $\mu_n^\omega(C)$ proves to be a "statistical picture" of $\mu(C)$ approximating $\mu(C)$ as the sample size n tends to infinity. Now the interesting question is, whether for a given subsystem $\mathcal{C}$ of $\mathcal{B}_k$, (1.3) holds uniformly in $\mathcal{C}$, i.e.

(1.4) $\lim_{n\to\infty}(\sup_{C\in\mathcal{C}}|\mu_n^\omega(C)-\mu(C)|) = 0$ $\mathbb{P}$-a.s.

Considering $\mathcal{C} = \mathcal{B}_k$ and $C(\omega) := \{\xi_i(\omega): i\in\mathbb{N}\}$ one obtains a set $C\in\mathcal{C}$ such that (1.4) would imply $\mu(C(\omega)) = 1$, i.e. μ must be necessarily discrete. On the other hand, if μ is supposed to be a discrete p.d. on $\mathcal{B}_k$, i.e. $\mu = \sum_{i\in T} m_i\varepsilon_{x_i}$, $x_i\in\mathbb{R}^k$, $m_i>0$, $\sum_{i\in T} m_i=1$, $T\subset\mathbb{N}$, then, by the strong law of large numbers, for some $N_1\in\mathcal{F}$ with $\mathbb{P}(N_1) = 0$: $\mu_n^\omega(\{x_i\}) \xrightarrow[n\to\infty]{} \mu(\{x_i\})$ for all $\omega\notin N_1$ and $i\in T$. Since μ is concentrated on $D:=\{x_i: i\in T\}$ there exists a second $\mathbb{P}$-null set N_2 such that for all $\omega\notin N_2$ $\mu_n^\omega(A) = 0$ for every

$A \subset \mathbb{R}^k \setminus D$. Hence, for all $\omega \notin N_1 \cup N_2$, $(\mu_n^\omega)_{n\in\mathbb{N}}$ is a sequence of measures on $(D, \mathfrak{P}(D))$ converging pointwise to μ. Applying Scheffé's lemma one obtains $\lim_{n\to\infty}(\sup_{\Delta \in \mathfrak{P}(D)} |\mu_n^\omega(\Delta)-\mu(\Delta)|) = 0$ and therefore (1.4). Hence

(1.5) $\lim_{n\to\infty} (\sup_{C \in \mathfrak{B}_k} |\mu_n^\omega(C)-\mu(C)|) = 0 \quad \Leftrightarrow \quad \mu$ discrete.

This shows that in order to allow more general p.d. μ one has to confine (1.4) on proper subsystems $\mathcal{C}$ of $\mathfrak{B}_k$ and the questions which one might then consider are:

A Given an arbitrary p.d. μ on $\mathfrak{B}_k$, for which $\mathcal{C} \subset \mathfrak{B}_k$ is it true that $D_n^\mu(\mathcal{C},\omega) := \sup_{C\in\mathcal{C}} |\mu_n^\omega(C)-\mu(C)| \to 0$ $\mathbb{P}$-a.s. as $n \to \infty$?

B Given $\mathcal{C} \subset \mathfrak{B}_k$, under which conditions on $\mu|\mathfrak{B}_k$ does it follow that $D_n^\mu(\mathcal{C},\omega) \to 0$ $\mathbb{P}$-a.s. as $n \to \infty$?

Let us first recall some known results in this field. To this extend let $\mathcal{M}(\mathbb{R}^k, \mathcal{C}) := \{\mu \in \mathcal{M}_+^1(\mathbb{R}^k): \lim_{n\to\infty} D_n^\mu(\mathcal{C},\omega) = 0 \ \mathbb{P}\text{-a.s.}\}$, where $\mathcal{M}_+^1(\mathbb{R}^k)$ denotes the space of all p-measures on $\mathfrak{B}_k$.

(i) Let $\mathcal{C} = {}_{-\infty}\mathcal{J}_k := \{I \subset \mathbb{R}^k: I = I(\underline{t}) := (-\underline{\infty}, \underline{t}]\ ,\ \underline{t} \in \mathbb{R}^k\}$; then $\mathcal{M}(\mathbb{R}^k, {}_{-\infty}\mathcal{J}_k) = \mathcal{M}_+^1(\mathbb{R}^k)$ which means, in terms of empirical d.f.'s (cf. (1.2)), that for arbitrary d.f.'s F $\lim_{n\to\infty} (\sup_{\underline{t}\in\mathbb{R}^k} |F_n^\omega(\underline{t})-F(\underline{t})|) = 0$ $\mathbb{P}$-a.s.

For $k=1$ this is the classical Glivenko-Cantelli theorem proved first by Glivenko [8] for continuous F and by Cantelli [1] for arbitrary F. For $k \geq 1$ the result apparently goes back to Kiefer and Wolfowitz (cf. [10] and [21]); an interesting and rather direct proof using compactness arguments in the space of all monotone functions defined on $\mathbb{R}^k$ was given by Dehardt [2] .

(ii) Let $\mathcal{C} = \mathcal{J}_k := \{I \subset \mathbb{R}^k: I = \prod_{i=1}^k (a_i, b_i]\ ,\ -\infty \leq a_i < b_i < \infty,\ i=1,\dots,k\}$, then again as in (i) $\mathcal{M}(\mathbb{R}^k, \mathcal{J}_k) = \mathcal{M}_+^1(\mathbb{R}^k)$.

Analogous results are true for $\mathcal{C} = \mathcal{H}_k$, the class of all halfspaces in $\mathbb{R}^k$, and for $\mathcal{C} = \mathcal{H}_k(m)$ consisting of all sets in $\mathbb{R}^k$ which are intersections of at most m half-spaces (cf. Wolfowitz [20] ,[21] and R.R. Rao [12]). Since all these $\mathcal{C}$'s consist of convex sets one might be tempted to expect that **A** is also true for $\mathcal{C} = \mathcal{C}_k := \{C \in \mathfrak{B}_k: C \text{ convex}\}$, but the following counterexample (for k=2) shows that this is not the case if $k \geq 2$:

Let S be the closed unit ball in $\mathbb{R}^2$ and let $\mu|\mathfrak{B}_2$ be the uniform distribution on the boundary ∂S of S. Then, if $(\xi_i)_{i\in\mathbb{N}}$ is any sequence of i.i.d. random vectors on some p-space $(\Omega, \mathfrak{F}, \mathbb{P})$ with values in $\mathbb{R}^2$ and p.d. μ, one obtains for all $\omega_0 \in \Omega_0 := \{\omega\in\Omega: \xi_i(\omega) \in \partial S,\ i=1,2,\dots\}$ (note that $\mathbb{P}(\Omega_0)=1$): $\lim_{n\to\infty} D_n^\mu(\mathcal{C}_2, \omega_0) = 1$. In fact, putting $C = C(\omega_0) = \{\xi_i(\omega_0): i \in \mathbb{N}\} \cup S^o$ (S^o=interior of S), then $C \in \mathcal{C}_2$, $\mu_n^{\omega_0}(C) = 1$ for all $n \in \mathbb{N}$, and (since μ is nonatomic) $\mu(C) = 0$.

Concerning question [B] the most well known answer is due to R.R. Rao [12] :

(iii) Let $\mathcal{C} = \mathcal{C}_k$ and suppose that $\mu | \mathcal{B}_k$ fulfills the following condition (+):

(+) $\sup_{C \in \mathcal{C}_k} \mu_c(\partial C) = 0$, where μ_c is the nonatomic part of μ and ∂C denotes the boundary of C.

Then $D_n^{\mu}(\mathcal{C}_k,\omega) \to 0$ $\mathbb{P}$-a.s. as $n \to \infty$.

As to other examples concerning [A] and [B] the reader is referred to the authors' survey [7] . It is the aim of the present paper to describe a rather general technique for proving theorems of the above kind. As to another method based on Lemma 2 in [6] which also yields (i)-(iii) above we refer to Krickeberg [11]. Our technique relies on the concept of uniformity classes as introduced in [5] and further developed by the second author (cf. [13], [14] and [15]). Into this direction goes also a forthcoming paper by Topsøe [18].

2. Uniformity classes. Let $(X, \mathcal{B})$ be a measurable space and denote with $\mathcal{M}_+(X)$ the set of all finite measures on $\mathcal{B}$. If (μ_α) is a net in $\mathcal{M}_+(X)$, $\mathcal{C}$ a subsystem of $\mathcal{B}$, and $\mu \in \mathcal{M}_+(X)$, the question arises about sufficient conditions for the validity of $\lim_\alpha (\sup_{C \in \mathcal{C}} |\mu_\alpha(C)-\mu(C)|) = 0$. Apparently, in this case it is necessary that in advance (μ_α) should approximate μ in an appropriate sense. If one is looking at the concept of weak convergence of measures in a metric space $(X, \mathcal{B})$ with $\mathcal{B} = \mathcal{B}(X)$ being the σ-algebra of Borel sets in X one knows from the Portmanteau theorem ([16], Theorem 8.1) that weak convergence of (μ_α) to μ is equivalent with setwise convergence of (μ_α) to μ on the field $\mathcal{A}_\mu := \{A \in \mathcal{B}(X): \mu(\partial A) = 0\}$. Hence in this context the approximation of μ by (μ_α) w.r.t. the topology of weak convergence may be described by setwise convergence of (μ_α) to μ on a suitably choosen sub-field $\mathcal{A}$ of $\mathcal{B}$. Concerning the above question this leads for arbitrary measurable spaces to the following definition.

2.1 Definition. Let $(X, \mathcal{B})$ be a measurable space and $\mu \in \mathcal{M}_+(X)$. Given a sub-field $\mathcal{A}$ of $\mathcal{B}$ a subsystem $\mathcal{C}$ of $\mathcal{B}$ is called a $(\mu, \mathcal{A})$-uniformity class iff $\lim_\alpha (\sup_{C \in \mathcal{C}} |\mu_\alpha(C)-\mu(C)|) = 0$ for any net (μ_α) in $\mathcal{M}_+(X)$ with $\lim_\alpha \mu_\alpha(A) = \mu(A)$ for all $A \in \mathcal{A}$.

The following theorem will give a useful characterization of $(\mu, \mathcal{A})$-uniformity classes. To state it, let $\Pi(\mathcal{A})$ denote the class of all finite partitions of X into $\mathcal{A}$-sets; $\pi^1 \in \Pi(\mathcal{A})$ is said to be a refinement of $\pi^2 \in \Pi(\mathcal{A})$ $[\pi^2 < \pi^1]$ iff $\pi^2 \subset \alpha(\pi^1)$, where $\alpha(\pi^1)$ denotes the field spanned by the elements of π^1. For every finite subcollection $\{\pi^1,\dots,\pi^m\}$ of $\Pi(\mathcal{A})$ let $\pi := \bigvee_{i=1}^m \pi^i$ be the common refinement of $\{\pi^1,\dots,\pi^m\}$. Since $\pi \in \Pi(\mathcal{A})$, $(\Pi(\mathcal{A}), <)$ turns out to be a directed system. In the sequel the elements of $\pi \in \Pi(\mathcal{A})$ are always assumed to be nonvoid. For $C \in \mathcal{C}$ and $\pi \in \Pi(\mathcal{A})$ let the "π-boundary" of C be defined by $\partial_\pi C := \bigcup \{A \in \pi: A \cap C \neq \emptyset \neq A \cap \complement C\}$ (where $\complement C$ denotes the complement of C); note that $\partial_{\pi^1} C \subset \partial_{\pi^2} C$, if $\pi^2 < \pi^1$. Now the following theorem holds ([13], Theorem 1.2).

2.2 Theorem. In the above notation $\mathcal{C} \subset \mathcal{B}$ is a $(\mu, \mathcal{A})$-uniformity class iff

(2.3) for every $\varepsilon > 0$ there exists $\pi = \pi(\varepsilon) \in \Pi(\mathcal{A})$ such that $\sup_{C \in \mathcal{C}} \mu(\partial_\pi C) \leq \varepsilon$.

Proof. To prove the sufficiency of (2.3), let $\varepsilon > 0$ be given. Then by assumption there exists $\pi = \pi(\varepsilon) \in \Pi(\mathcal{A})$ with $\sup_{C \in \mathcal{C}} \mu(\partial_\pi C) \leq \varepsilon/2$. For every $C \in \mathcal{C}$ let $C^- = C^-(\pi) := \bigcup\{A \in \pi : A \subset C\}$ and $C^+ = C^+(\pi) := \bigcup\{A \in \pi : A \cap C \neq \emptyset\}$; then

(i) $C^- \in \alpha(\pi), C^+ \in \alpha(\pi)$; (ii) $C^- \subset C \subset C^+$; (iii) $C^+ \setminus C^- = \partial_\pi C$.

Now, let (μ_α) be any net in $\mathcal{M}_+(X)$ with $\lim_\alpha \mu_\alpha(A) = \mu(A)$ for all $A \in \mathcal{A}$. Since $\alpha(\pi) \subset \mathcal{A}$ and as $\alpha(\pi)$ is finite, there exists α_0 such that $\sup_{A \in \alpha(\pi)} |\mu_\alpha(A) - \mu(A)| \leq \varepsilon/2$ for all $\alpha > \alpha_0$. Hence one obtains from (i) for each $\alpha > \alpha_0$:

$\sup_{C \in \mathcal{C}} |\mu_\alpha(C^+) - \mu(C^+)| \leq \varepsilon/2$ and $\sup_{C \in \mathcal{C}} |\mu_\alpha(C^-) - \mu(C^-)| \leq \varepsilon/2$, and therefore according to (ii) and (iii) $\mu_\alpha(C) - \mu(C) \leq \mu_\alpha(C^+) - \mu(C^-) \leq |\mu_\alpha(C^+) - \mu(C^+)| + \varepsilon/2 \leq \varepsilon$ and $\mu_\alpha(C) - \mu(C) \geq \mu_\alpha(C^-) - \mu(C^+) \geq -|\mu_\alpha(C^-) - \mu(C^-)| - (\mu(C^+) - \mu(C^-)) \geq -\varepsilon/2 - \varepsilon/2 = -\varepsilon$, i.e. $\sup_{C \in \mathcal{C}} |\mu_\alpha(C) - \mu(C)| \leq \varepsilon$ for all $\alpha > \alpha_0$.

To prove the necessity of (2.3), suppose that there exists $\varepsilon_1 > 0$ such that for all $\pi \in \Pi(\mathcal{A})$ there is a $C_\pi \in \mathcal{C}$ with $\mu(\partial_\pi C_\pi) > \varepsilon_1$. From this we claim: For each $\pi \in \Pi(\mathcal{A})$ there exists $\mu_\pi \in \mathcal{M}_+(X)$ with the following properties

(*) $\mu_\pi(A) = \mu(A)$ for all $A \in \pi$ and (**) $|\mu_\pi(C_\pi) - \mu(C_\pi)| \geq \varepsilon_1/2$.

In fact, since $\mu(\partial_\pi C_\pi) > \varepsilon_1$ at least one of the following two cases will necessarily occur:

(a) $\mu(\partial_\pi C_\pi \cap C_\pi) > \varepsilon_1/2$ or (b) $\mu(\partial_\pi C_\pi \setminus C_\pi) > \varepsilon_1/2$.

In case (b) the measure μ_π fulfilling (*) and (**) is obtained in the following way ("moving the mass of $\partial_\pi C_\pi \setminus C_\pi$ into C_π"): If $\partial_\pi C_\pi = \sum_i \Delta_i$ and x_i is an arbitrary point in $\Delta_i \cap C_\pi$, we define $\mu_\pi := \mathrm{rest}_{\complement(\partial_\pi C_\pi \setminus C_\pi)}\mu + \sum_i \mu(\Delta_i \setminus C_\pi)\varepsilon_{x_i}$. Then $\mu_\pi(A) = \mu(A \cap \complement(\partial_\pi C_\pi \setminus C_\pi)) + \sum_i \mu(\Delta_i \setminus C_\pi)\varepsilon_{x_i}(A) = \mu(A)$ for all $A \in \pi$ and $\mu_\pi(C_\pi) = \mu(C_\pi \cap \complement(\partial_\pi C_\pi \setminus C_\pi)) + \mu(\partial_\pi C_\pi \setminus C_\pi) = \mu(C_\pi) + \mu(\partial_\pi C_\pi \setminus C_\pi)$, i.e. $\mu_\pi(C_\pi) - \mu(C_\pi) = \mu(\partial_\pi C_\pi \setminus C_\pi) > \varepsilon_1/2$. Furthermore, $\lim_\pi \mu_\pi(A) = \mu(A)$ for all $A \in \mathcal{A}$. To this extent let $A_0 \in \mathcal{A}$ be arbitrary and suppose w.l.o.g. $A_0 \neq \emptyset, X$. Put $\pi_0 := \{A_0, \complement A_0\}$ and let $\pi > \pi_0$; then it is easy to see that $\mu_\pi(A_0) = \mu(A_0)$ which proves that (μ_π) is a net in $\mathcal{M}_+(X)$ converging setwise to μ on $\mathcal{A}$. By assumption this implies $\lim_\pi (\sup_{C \in \mathcal{C}} |\mu_\pi(C) - \mu(C)|) = 0$ which contradicts (**). In case (a) the proof runs analogously "moving the mass of $\partial_\pi C_\pi \cap C_\pi$ into $\complement C_\pi$". □

For later applications (cf. 3.1 below) the following lemma will be crucial.

2.4 Lemma. Let $\mathcal{C} \subset \mathcal{B}$ be a $(\mu, \mathcal{A})$-uniformity class. Then there exists a countable sub-field $\mathcal{A}_o$ of $\mathcal{A}$ such that $\mathcal{C}$ is even a $(\mu, \mathcal{A}_o)$-uniformity class.

Proof. Apply 2.2 to obtain for every $\varepsilon_n = 1/n$, $n \in \mathbb{N}$, a $\pi_n \in \Pi(\mathcal{A})$ such that $\sup_{C \in \mathcal{C}} \mu(\partial_{\pi_n} C) \leq 1/n$. Let $\mathcal{A}_o := \alpha(\{A: A \in \pi_n \text{ for some } n \in \mathbb{N}\})$; then $\mathcal{A}_o$ is a countable sub-field of $\mathcal{A}$. To show that $\mathcal{C}$ is a $(\mu, \mathcal{A}_o)$-uniformity class, for any $\varepsilon > 0$ let $n_o \in \mathbb{N}$ be such that $1/n_o \leq \varepsilon$; as $\pi_{n_o} \in \Pi(\mathcal{A}_o)$ and $\sup_{C \in \mathcal{C}} \mu(\partial_{\pi_{n_o}} C) \leq 1/n_o \leq \varepsilon$, the assertion follows from 2.2. □

For technical reasons we need the following "decomposition lemma".

2.5 Lemma. Let $\mu \in \mathcal{M}_+(X)$ and let $(\mu_i)_{i \geq 0}$ be a sequence in $\mathcal{M}_+(X)$ with $\sum_{i \geq 0} \mu_i = \mu$. Then $\mathcal{C}$ is a $(\mu, \mathcal{A})$-uniformity class if $\mathcal{C}$ is a $(\mu_i, \mathcal{A})$-uniformity class for every $i \geq 0$.

Proof. Given $\varepsilon > 0$ it follows by the finiteness of $\mu(X)$ that there exists $i_o = i_o(\varepsilon) \in \mathbb{N}$ with $\sum_{i > i_o} \mu_i(X) < \varepsilon/2$. Since by assumption $\mathcal{C}$ is a $(\mu_i, \mathcal{A})$-uniformity class for $i = 0, 1, \ldots, i_o$, it follows from 2.2 that there exist $\pi_i \in \Pi(\mathcal{A})$ with $\sup_{C \in \mathcal{C}} \mu_i(\partial_{\pi_i} C) \leq \varepsilon[2(i_o+1)]^{-1}$, $i = 0, 1, \ldots, i_o$. Let $\pi = \bigvee_{i=0}^{i_o}$ be the common refinement of $\pi_o, \pi_1, \ldots, \pi_{i_o}$; then $\pi \in \Pi(\mathcal{A})$ and $\sup_{C \in \mathcal{C}} \sum_{i=0}^{i_o} \mu_i(\partial_\pi C) \leq \sum_{i=0}^{i_o} \sup_{C \in \mathcal{C}} \mu_i(\partial_{\pi_i} C) \leq \varepsilon/2$ and therefore $\sup_{C \in \mathcal{C}} \mu(\partial_\pi C) \leq \varepsilon$. The assertion now follows from 2.2 again. □

2.6 Remark (cf. (1.5)). Let $\mu \in \mathcal{M}_+(X)$ be discrete and $\mathcal{A}$ a sub-field of $\mathcal{B}$ containing all singletons $\{x\}$, $x \in X$. Then $\mathcal{B}$ is a $(\mu, \mathcal{A})$-uniformity class.

Proof. Since μ is discrete, $\mu = \sum_{i \in T} \mu_i$, $T \subset \mathbb{N}$, with $\mu_i = \mu(\{x_i\}) \cdot \varepsilon_{x_i}$, $x_i \in X$. Hence, by 2.5, it suffices to show that $\mathcal{B}$ is a $(\mu_i, \mathcal{A})$-uniformity class for every $i \in T$. This follows at once from 2.2 choosing for $i \in T$ $\pi_i := \{\{x_i\}, \complement\{x_i\}\} \in \Pi(\mathcal{A})$, and noticing that $\partial_{\pi_i} C \subset \complement\{x_i\}$ for all $C \in \mathcal{C}$, i.e. $\sup_{C \in \mathcal{C}} \mu_i(\partial_{\pi_i} C) = 0$. □

Given an arbitrary $\mu \in \mathcal{M}_+(X)$ μ may be decomposed into $\mu = \mu_d + \mu_c$ where μ_d is discrete and μ_c is the nonatomic part of μ (i.e. $\mu_c(\{x\}) = 0$ for all $x \in X$). It follows from 2.5 and 2.6 that in order to show that $\mathcal{C}$ is a $(\mu, \mathcal{A})$-uniformity class (where $\mathcal{A}$ is supposed to have the property stated in 2.6) it suffices to prove that $\mathcal{C}$ is a $(\mu_c, \mathcal{A})$-uniformity class, i.e. in the sequel we may and do assume that $\mu = \mu_c$.

Concerning [A] in the introduction we are interested into those $\mathcal{C} \subset \mathcal{B}$ which are $(\mu, \mathcal{A})$-uniformity classes for *any* $\mu \in \mathcal{M}_+(X)$.

2.7 Definition. $\mathcal{C} \subset \mathcal{B}$ is called an ideal $(X, \mathcal{A})$-uniformity class iff $\mathcal{C}$ is a $(\mu, \mathcal{A})$-uniformity class for every $\mu \in \mathcal{M}_+(X)$.

As a trivial fact we remark that all finite $\mathcal{C} \subset \mathcal{A}$ are ideal $(X,\mathcal{A})$-uniformity classes.

Now our first aim is to prove that for $(X,\mathfrak{B}) = (\mathbb{R}^k, \mathfrak{B}_k)$, $k \geq 1$, and $\mathcal{A} = \mathfrak{B} = \mathfrak{B}_k$ the class $\mathcal{C} = {}_{-\infty}\mathfrak{J}_k$ (cf. 1. (i)) is an ideal $(\mathbb{R}^k, \mathfrak{B}_k)$-uniformity class. To this extent we need the following lemma which proves to be useful also in other cases.

<u>2.8 Lemma.</u> Let $(X_i, \mathfrak{B}_i)$, $i=1,2$, be measurable spaces and let $f: X_1 \to X_2$ be $\mathfrak{B}_1, \mathfrak{B}_2$-measurable. If $\mathcal{C}_2 \subset \mathfrak{B}_2$ is an ideal $(X_2, \mathfrak{B}_2)$-uniformity class, then $\mathcal{C}_1 = f^{-1}(\mathcal{C}_2) := \{f^{-1}(C_2): C_2 \in \mathcal{C}_2\}$ is an ideal $(X_1, \mathfrak{B}_1)$-uniformity class.

<u>Proof.</u> Let $\mu_1 \in \mathcal{M}_+(X_1)$ be arbitrary and consider a net (μ_α) in $\mathcal{M}_+(X_1)$ with $\lim_\alpha \mu_\alpha(A_1) = \mu_1(A_1)$ for all $A_1 \in \mathfrak{B}_1$. Put $\nu_\alpha := f\mu_\alpha$ (the image measure of μ_α under f) and $\mu_2 := f\mu_1$; then $\mu_2 \in \mathcal{M}_+(X_2)$ and (ν_α) is a net in $\mathcal{M}_+(X_2)$ with $\lim_\alpha \nu_\alpha(A_2) = \mu_2(A_2)$ for all $A_2 \in \mathfrak{B}_2$. Hence, by assumption $\lim_\alpha (\sup_{C_2 \in \mathcal{C}_2} |\nu_\alpha(C_2) - \mu_2(C_2)|) = 0$ which implies (by the definition of $\mathcal{C}_2$, ν_α and μ_2, respectively) that $\lim_\alpha (\sup_{C_1 \in \mathcal{C}_1} |\mu_\alpha(C_1) - \mu(C_1)|) = 0$. □

<u>2.9 Theorem.</u> $\mathcal{C} = {}_{-\infty}\mathfrak{J}_k$, $k \geq 1$, is an ideal $(\mathbb{R}^k, \mathfrak{B}_k)$-uniformity class.

<u>Proof.</u> (By induction on k). <u>k=1</u>: Let $\mu \in \mathcal{M}_+(\mathbb{R})$ and suppose w.l.o.g. that $\mu = \mu_c$. Then, for given $\varepsilon > 0$, choose $-\infty < t_1 < t_2 < \ldots < t_n < \infty$ such that $\mu(I(t_1)) < \varepsilon$, $\mu(\complement I(t_n)) < \varepsilon$ and $\mu(I(t_l) \setminus I(t_{l-1})) < \varepsilon$, $l=2,\ldots,n$. Put $\pi := \pi(\varepsilon)$: $\{I(t_1), \complement I(t_n), I(t_l) \setminus I(t_{l-1}): l=2,\ldots,n\}$; then $\pi \in \Pi(\mathfrak{B})$ and $\sup_{I \in {}_{-\infty}\mathfrak{J}_1} \mu(\partial_\pi I) \leq \varepsilon$.

<u>k≥2</u>: For given $\mu \in \mathcal{M}_+(\mathbb{R}^k)$ denote with $\mu^{\{s\}} = p_s\mu$ the s-th (one-dimensional) marginal measure pertaining to μ, $s=1,2,\ldots,k$. Then, for every s, there exist at most countably many $x_j^s \in \mathbb{R}$, $j \in T_s \subset \mathbb{N}$, with $\mu^{\{s\}}(\{x_j^s\}) > 0$. Let μ_j^s be the restriction of μ onto $H_j^s := \mathbb{R} \times \mathbb{R} \times \ldots \times \mathbb{R} \times \{x_j^s\} \times \mathbb{R} \times \ldots \times \mathbb{R}$, $1 \leq s \leq k$, $j \in T_s$, and put $H := \bigcup_{s=1}^k \bigcup_{j \in T_s} H_j^s$. Then there exists a partition of H into disjoint sets B_i, $i \in \mathbb{N}$, such that every B_i is contained in some H_j^s. Put $\mu_i := \mathrm{rest}_{B_i}\mu$ and $\mu_o := \mathrm{rest}_{\complement H}\mu$; then $\mu = \sum_{i \geq 0} \mu_i$, whence by 2.5 it suffices to show that ${}_{-\infty}\mathfrak{J}_k$ is a $(\mu_i, \mathfrak{B}_k)$-uniformity class for each $i \geq 0$.

$i \geq 1$: Let H_j^s be chosen so that $B_i \subset H_j^s$ and define $f_j^s: H_j^s \to \mathbb{R}^{k-1}$ by $f_j^s(x_1,\ldots,x_{s-1},x_j^s,x_{s+1},\ldots,x_k) := (x_1,\ldots,x_{s-1},x_{s+1},\ldots,x_k)$. Clearly f_j^s is $H_j^s \cap \mathfrak{B}_k, \mathfrak{B}_{k-1}$-measurable with $(f_j^s)^{-1}({}_{-\infty}\mathfrak{J}_{k-1}) = H_j^s \cap {}_{-\infty}\mathfrak{J}_k = \{H_j^s \cap I: I \in {}_{-\infty}\mathfrak{J}_k\}$. By the induction hypothesis ${}_{-\infty}\mathfrak{J}_{k-1}$ is an ideal $(\mathbb{R}^{k-1}, \mathfrak{B}_{k-1})$-uniformity class and therefore by 2.8 $H_j^s \cap {}_{-\infty}\mathfrak{J}_k$ is an ideal $(H_j^s, H_j^s \cap \mathfrak{B}_k)$-uniformity class. Considering μ_i as a measure on $(H_j^s, H_j^s \cap \mathfrak{B}_k)$ it follows from 2.2 that for every $\varepsilon > 0$ there exists a partition $\pi^o = \pi^o(\varepsilon) \in \Pi(H_j^s \cap \mathfrak{B}_k)$ with $\sup_{I \in H_j^s \cap {}_{-\infty}\mathfrak{J}_k} \mu_i(\partial_{\pi^o} I) \leq \varepsilon$. Let $\pi := \{\pi^o, \complement H_j^s\}$; then $\pi \in \Pi(\mathfrak{B}_k)$ and

$\sup_{I \in {}_{-\infty}\mathcal{J}_k} \mu_i(\partial_\pi I) = \sup_{I \in H_j^s \cap {}_{-\infty}\mathcal{J}_k} \mu_i(\partial_{\pi_0} I) \leq \varepsilon$, which proves that ${}_{-\infty}\mathcal{J}_k$ is a $(\mu_i, \mathfrak{B}_k)$-uniformity class for every $i \geq 1$.

i=0: For any $\varepsilon > 0$ choose $-\infty < t_1^s < t_2^s < \ldots < t_{n_s}^s < \infty$, $s=1,\ldots,k$, in such a way that

$\max\{\mu_0^{\{s\}}(A) : A \in \pi^s\} \leq \varepsilon/k$, where $\pi^s := \{I(t_1^s), \complement I(t_{n_s}^s), I(t_l^s) \setminus I(t_{l-1}^s), l=2,\ldots,n_s\}$.

Let $\pi \in \Pi(\mathfrak{B}_k)$ be the partition whose elements consist of boxes with sides pertaining to π^s, $s=1,\ldots,k$; we claim: $\sup_{I \in {}_{-\infty}\mathcal{J}_k} \mu_0(\partial_\pi I) \leq \varepsilon$. Indeed, given $I = I(\underline{t}) = (-\underline{\infty}, \underline{t}]$, then $t_s \in A_s$ for exactly one $A_s \in \pi^s$, $s=1,\ldots,k$. It follows $\partial_\pi I \subset \bigcup_{s=1}^k \mathbb{R} \times \ldots \times \mathbb{R} \times A_s \times \mathbb{R} \times \ldots \times \mathbb{R}$, and therefore $\mu_0(\partial_\pi I) \leq \sum_{s=1}^k \mu^{\{s\}}(A_s) \leq k \cdot \varepsilon/k = \varepsilon$. This proves 2.9. □

With the aid of 2.9 and 2.8 one can prove uniformity properties for a whole variety of classes $\mathcal{C}_1 \subset \mathfrak{B}_1$. Let us confine here to prove the following corollary.

<u>2.10 Corollary.</u> Let $(X, \|\cdot\|)$ be a normed space with $\mathfrak{B} = \mathfrak{B}(X)$ the Borel σ-field in X. Fix an arbitrary $x_0 \in X$ and denote with $\mathcal{C}(x_0)$ the class of all closed balls in X with center x_0. Then $\mathcal{C}(x_0)$ is an ideal $(X, \mathfrak{B}(X))$-uniformity class.

<u>Proof.</u> Apply 2.8 with $(X_1, \mathfrak{B}_1) = (X, \mathfrak{B}(X))$, $(X_2, \mathfrak{B}_2) = (\mathbb{R}, \mathfrak{B})$ and $f = f_{x_0} : X \to \mathbb{R}$, defined by $f(x) := \|x - x_0\|$; then f is $\mathfrak{B}(X), \mathfrak{B}$-measurable and $\mathcal{C}(x_0) = f^{-1}({}_{-\infty}\mathcal{J}_1)$; hence the assertion follows immediately from 2.9 and 2.8. □

<u>2.11 Remark.</u> It was shown by Elker [3] that for $X = \mathbb{R}^k$ 2.10 even holds if one replaces $\mathcal{C}(x_0)$ by the class of <u>all</u> closed balls (with arbitrary center). On the other hand it follows from the examples in [17] that the same is no longer true for arbitrary (infinite-dimensional) spaces $(X, \|\cdot\|)$.

As a further corollary we obtain immediately from 2.9 the following result for $\mathcal{C} = \mathcal{J}_k$ (cf. 1 (ii)).

<u>2.12 Corollary.</u> $\mathcal{C} = \mathcal{J}_k$, $k \geq 1$, is an ideal $(\mathbb{R}^k, \mathfrak{B}_k)$-uniformity class.

Next we will face our attention on the class $\mathcal{C} = \mathcal{C}_k$ of all measurable convex sets in $\mathbb{R}^k$, $k \geq 2$. According to the example mentioned in the introduction one will expect that $\mathcal{C}_k$, $k \geq 2$, is not an ideal $(\mathbb{R}^k, \mathfrak{B}_k)$-uniformity class (cf. 2.17 below). The following theorem gives a sufficient condition on μ for $\mathcal{C}_k$ being a $(\mu, \alpha(\mathcal{C}_k))$-uniformity class (where $\alpha(\mathcal{C}_k)$ denotes the field spanned by $\mathcal{C}_k$).

<u>2.13 Theorem.</u> Suppose that $\mu \in \mathcal{M}_+(\mathbb{R}^k)$ fulfills (+): $\sup_{C \in \mathcal{C}_k} \mu_c(\partial C) = 0$. Then $\mathcal{C}_k$ is a $(\mu, \alpha(\mathcal{C}_k))$-uniformity class.

The proof of this theorem will be based on the so called "Blaschke selection theorem". To make some preliminary remarks in this context, for all $\rho > 0$ and $A \subset \mathbb{R}^k$ let $A^\rho := \{x \in \mathbb{R}^k:$

$\|x,A\| < \rho\}$, where $\|x,A\| := \inf_{a\in A} \|x-a\|$, $\|\cdot\|$ being the Euclidean norm in $\mathbb{R}^k$. Further, let $\partial_\rho A:=\{x\in\mathbb{R}^k: \|x,A\| < \rho$ and $\|x,\complement A\| < \rho\}$; note that A^ρ as well as $\partial_\rho A$ are open subsets of $\mathbb{R}^k$, hence $A_\rho := A^\rho \setminus \partial_\rho A \in \mathcal{B}_k$. If A,B are nonvoid subsets of $\mathbb{R}^k$, put $d(A,B):= \inf\{\rho > 0: A\subset B^\rho$ and $B\subset A^\rho\}$. Then d proves to be a metric on the space of all nonvoid, bounded, closed and convex sets in $\mathbb{R}^k$, the so called Hausdorff metric. Now, Blaschke's theorem is as following.

<u>2.14 Lemma</u> (cf. [19], p. 47). Let $\mathcal{C}_k^r$ be the class of all nonvoid, closed and convex subsets of $B(\underline{0},r):=\{x\in\mathbb{R}^k: \|x\| \leqq r\}$. Then $(\mathcal{C}_k^r,d)$ is a compact metric space.

<u>2.15 Remark.</u> The following statements are easy to prove:

(i) $C\in\mathcal{C}_k^r$ implies that C^ρ as well as C_ρ are convex, hence in $\mathcal{C}_k$;

(ii) $C,C'\in\mathcal{C}_k^r$ and $d(C,C') < \delta$ implies that $C_\delta\subset C'\subset C^\delta$.

Now we are in the position to give the <u>proof of 2.13</u>. According to the remarks preceding 2.7 we may and do assume w.l.o.g. that $\mu = \mu_c$. Let $\varepsilon > 0$ be given and choose $r > 0$ so that $\mu(\complement B(\underline{0},r)) \leqq \varepsilon/2$. It follows from the assumption (+) that for any $C\in\mathcal{C}_k$, $\emptyset \neq C\subset B(\underline{0},r)$, there exists a $\delta = \delta(\varepsilon,C) > 0$ s.t.

(*) $\mu(\partial_\delta C) \leqq \varepsilon/2$ (note that $\bigcap_{\delta>0} \partial_\delta C = \partial C$ and $\partial_\delta C \downarrow \partial C$).

Let $U_\delta(C):=\{C'\in\mathcal{C}_k^r: d(C,C') < \delta\}$ with $C\in\mathcal{C}_k^r$ and $\delta = \delta(\varepsilon,C)$; then $\bigcup_{C\in\mathcal{C}_k^r} U_\delta(C)$ is an open covering of $(\mathcal{C}_k^r,d)$. Hence by 2.14 there exist $C_1,\dots,C_n\in\mathcal{C}_k^r$ s.t. for any $C\in\mathcal{C}_k^r$ there is an $i\in\{1,\dots,n\}$ with $d(C,C_i) < \delta_i = \delta(\varepsilon,C_i)$. Writing $\{C_i^{\delta_i}: i=1,\dots,n\}\cup\{(C_i)_{\delta_i}: i=1,\dots,n\}\cup\{\complement B(\underline{0},r)\} =: \{A_j: j=1,\dots,m\}$, then by 2.15 (i) $\pi :=$ $\{\bigcap_{j\in T} A_j \cap \bigcap_{j\in\{1,\dots,m\}\setminus T} \complement A_j: T\subset\{1,\dots,m\}\}$ defines a finite partition of $\mathbb{R}^k$ which belongs to $\pi(\alpha(\mathcal{C}_k))$.

1. We claim: $\sup_{C\in\mathcal{C}_k,\ C \text{ closed}} \mu(\partial_\pi C) \leqq \varepsilon$: Since $\partial_\pi C\subset \partial_\pi(C\cap B(\underline{0},r)) \cup \complement B(\underline{0},r)$ and as $\mu(\complement B(\underline{0},r)) \leqq \varepsilon/2$, it suffices to show that $\mu(\partial_\pi C) \leqq \varepsilon/2$ for all $C\in\mathcal{C}_k^r$. Consider an arbitrary $C\in\mathcal{C}_k^r$ and choose $i\in\{1,\dots,n\}$ s.t. $d(C,C_i) < \delta_i = \delta(\varepsilon,C_i)$; then by 2.15 (ii) $(C_i)_{\delta_i}\subset C\subset C_i^{\delta_i}$. According to (*) it suffices to show that $\partial_\pi C\subset\partial_{\delta_i} C_i$. To this extent consider an arbitrary $A\in\pi$ with $A\subset\partial_\pi C$. Then $A=\bigcap_{j\in T} A_j \cap \bigcap_{j\in\{1,\dots,m\}\setminus T} \complement A_j$ for some $T\subset\{1,\dots,m\}$, and since $A\cap C \neq \emptyset$ together with $C\subset C_i^{\delta_i}$ imply $A\cap C_i^{\delta_i} \neq \emptyset$, one obtains $C_i^{\delta_i} = A_j$ for some $j\in T$, i.e. $A\subset C_i^{\delta_i}$. It remains to show that $A\cap(C_i)_{\delta_i}=\emptyset$. Suppose $A\cap(C_i)_{\delta_i}\neq\emptyset$; then $A\subset(C_i)_{\delta_i}\subset C$ which contradicts $A\cap\complement C \neq \emptyset$.

2. It follows from 1. that $\mathcal{C}_k^c:=\{C\in\mathcal{C}_k: C \text{ closed}\}$ is a $(\mu,\alpha(\mathcal{C}_k))$-uniformity class, and therefore $\lim_\alpha (\sup_{C\in\mathcal{C}_k^c} |\mu_\alpha(C)-\mu(C)|) = 0$ for any net in $\mathcal{M}_+(\mathbb{R}^k)$ with $\lim_\alpha \mu_\alpha(C_o) = \mu(C_o)$ for all $C_o\in\alpha(\mathcal{C}_k)$. Hence the assertion in 2.13 will be proved by showing that

(**) $\sup_{C\in\mathcal{C}_k} |\mu_\alpha(C)-\mu(C)| = \sup_{C\in\mathcal{C}_k^c} |\mu_\alpha(C)-\mu(C)|$.

As to (**), suppose that $|\mu_\alpha(C)-\mu(C)| = \mu_\alpha(C)-\mu(C)$; then (denoting with C^c the closure of C) $|\mu_\alpha(C)-\mu(C)| \leq |\mu_\alpha(C^c)-\mu(C^c)|$ (note that by assumption $\mu(\partial C)=0$). On the other hand, if $|\mu_\alpha(C)-\mu(C)| = \mu(C)-\mu_\alpha(C)$, choose for any $\varepsilon>0$ $C_1\in\mathcal{C}_k^c$ s.t. $C_1\subset C$ and $\mu(C\setminus C_1)\leq\varepsilon$; then $|\mu_\alpha(C)-\mu(C)| \leq |\mu_\alpha(C_1)-\mu(C_1)| + \varepsilon$, whence $\sup_{C\in\mathcal{C}_k} |\mu_\alpha(C)-\mu(C)| \leq \sup_{C\in\mathcal{C}_k^c} |\mu_\alpha(C)-\mu(C)| + \varepsilon$ which implies (**). □

The following result due to J. Elker shows how far away the assumption (+) in 2.13 is from being a necessary condition on μ.

2.16 Theorem. Let $\mu\in\mathcal{M}_+(\mathbb{R}^k)$; then

(+) $\sup_{C\in\mathcal{C}_k} \mu_c(\partial C) = 0$ if and only if

(i) $\mu_c(H) = 0$ for all hyperplanes H in $\mathbb{R}^k$ and

(ii) $\mathcal{C}_k$ is a $(\mu,\alpha(\mathcal{C}_k))$-uniformity class.

Proof. According to 2.13 it remains to show that (i) and (ii) are sufficient for (+). Again we may assume w.l.o.g. that $\mu = \mu_c$. Suppose that there exists $C_o\in\mathcal{C}_k$ with $\mu(\partial C_o)>0$; then, by (ii), it follows from 2.2 that for every $0<\varepsilon<\mu(\partial C_o)$ there exists $\pi = \pi(\varepsilon)\in\Pi(\alpha(\mathcal{C}_k))$ s.t. (*) $\sup_{C\in\mathcal{C}_k} \mu(\partial_\pi C)\leq\varepsilon$. Put $\sigma := \{A\in\pi: \mu(A\cap\partial C_o)>0\}$ ($\neq\emptyset$, since $\mu(\partial C_o)>0$) and let x_A be an arbitrary point in $A\cap\partial C_o$. Then it follows from (i) that (**) $\mu(\partial D)=0$ where D is the convex hull of $\{x_A: A\in\sigma\}$. Furthermore, since $\mu(A\cap\partial C_o)>0$ for all $A\in\sigma$ and μ is nonatomic there exist $x_A'\in A\cap\partial C_o$, $x_A'\neq x_A$, and $x_A'\notin\partial D$ (note that $A\cap\partial C_o\subset\partial D$ together with (**) would imply that $\mu(A\cap\partial C_o)=0$). Now, $D\subset C_o^c$ implies $D^o\subset(C_o^c)^o = C_o^o$ (where A^o denotes the interior of A), whence $\partial C_o\cap D^o=\emptyset$ and therefore $x_A'\in\complement D$ for all $A\in\sigma$. Hence $\bigcup\{A: A\in\sigma\}\subset\partial_\pi D$ which implies $\varepsilon<\mu(\partial C_o) = \mu(\partial C_o\cap\bigcup\{A: A\in\sigma\})\leq\mu(\bigcup\{A: A\in\sigma\})\leq\mu(\partial_\pi D)$; as $D\in\mathcal{C}_k$ this contradicts (*) and the theorem is proved. □

2.17 Remark. Note that in order to show the necessity of (+) we only used the fact that $\pi\in\Pi(\mathcal{B}_k)$ and not even $\pi\in\Pi(\alpha(\mathcal{C}_k))$. Thus 2.16 still remains true if one replaces $\alpha(\mathcal{C}_k)$ in (ii) by $\mathcal{B}_k$, showing that $\mathcal{C}_k$, $k\geq 2$, is not an ideal $(\mathbb{R}^k,\mathcal{B}_k)$-uniformity class. For, let μ be the uniform distribution on the boundary ∂S of the unit ball S in $\mathbb{R}^k$, $k\geq 2$. Then obviously (i) of 2.16 is fulfilled with $\mu_c(\partial S)=1>0$, hence $\mathcal{C}_k$ cannot be a $(\mu,\mathcal{B}_k)$-uniformity class.

From 2.13 we also obtain the following generalization of a result by Fabian [4] (Theorem 4.1):

2.18 Theorem. Let $\mu\in\mathcal{M}_+(\mathbb{R}^k)$ and let (μ_α) be a net in $\mathcal{M}_+(\mathbb{R}^k)$ with $\lim_\alpha \mu_\alpha(C)=\mu(C)$ for all $C\in\mathcal{C}_k$; then (+) $\sup_{C\in\mathcal{C}_k} \mu_c(\partial C) = 0$ implies $\lim_\alpha (\sup_{C\in\mathcal{C}_k} |\mu_\alpha(C)-\mu(C)|) = 0$.

Proof. $\lim_\alpha \mu_\alpha(C)=\mu(C)$ for all $C \in \mathcal{C}_k$ implies $\lim_\alpha \mu_\alpha(C_o)=\mu(C_o)$ for all $C_o \in \alpha(\mathcal{C}_k)$ (note that $\mathcal{C}_k$ is $\cap$-closed and $\mathbb{R}^k \in \mathcal{C}_k$). Hence the assertion is an immediate consequence of 2.13. □

2.19 Corollary. Let $\mu \in \mathcal{M}_+(\mathbb{R}^k)$ and (μ_α) be a net in $\mathcal{M}_+(\mathbb{R}^k)$ converging weakly to μ $(\mu_\alpha \xrightarrow[w]{} \mu)$; then

(++) $\sup_{C \in \mathcal{C}_k} \mu(\partial C)=0$ implies that $\lim_\alpha (\sup_{C \in \mathcal{C}_k} |\mu_\alpha(C)-\mu(C)|)=0.$

Proof. By the Portmanteau theorem $\mu_\alpha \xrightarrow[w]{} \mu$ iff $\lim_\alpha \mu_\alpha(A)=\mu(A)$ for all $A \in \mathcal{A}_\mu = \{A \in \mathfrak{B}_k: \mu(\partial A)=0\}$. Now, by (++), $\mathcal{C}_k \subset \mathcal{A}_\mu$, and the assertion follows from 2.18. □

The following lemma shows that (++) (and hence also (+)) is automatically satisfied whenever μ is absolutely continuous with respect to the product of k nonatomic Radon measures $\mu_i|\mathfrak{B}$ $(\mu << \bigotimes_{i=1}^k \mu_i)$; especially (++) holds true for any measure μ being absolutely continuous w.r.t. the k-dimensional Lebesgue measure.

2.20 Lemma. Let $\mu \in \mathcal{M}_+(\mathbb{R}^k)$ and suppose that there are nonatomic Radon measures $\mu_i|\mathfrak{B}$, $i=1,\dots,k$, such that $\mu << \bigotimes_{i=1}^k \mu_i =: \nu$; then (++) $\sup_{C \in \mathcal{C}_k} \mu(\partial C)=0$.

Proof. We must show that for any $\varepsilon > 0$ one has $\sup_{C \in \mathcal{C}_k} \mu(\partial C) \leqq \varepsilon$. First we can choose $K_k=K_k(\varepsilon):=[-a,a]^k$ in such a way that $\mu(\complement K_k) \leqq \varepsilon$, whence it remains to show that $\sup_{C \in K_k \cap \mathcal{C}_k} \nu(\partial C)=0$. W.l.o.g. we will consider the case k=2 and proceed inductively in the following way: In the first step we choose a subdivision of K_2 into nine subintervals of the form $I_i^1 = I_{j_i}^1 \times I_{k_i}^1$, $j_i,k_i=1,2,3$, $i=1,\dots,9$, where $I_{j_i}^1$ and $I_{k_i}^1$ are chosen so that $\mu_1(I_{j_i}^1) = \frac{1}{3}\mu_1([-a,a])$ and $\mu_2(I_{k_i}^1) = \frac{1}{3}\mu_2([-a,a])$, respectively, which is possible since μ_1 and μ_2 are supposed to be nonatomic (cf. [9] Halmos, p. 174). Thus $\nu(I_i^1)=\frac{\nu(K_2)}{9}$. Now, given $C \in K_2 \cap \mathcal{C}_2$ there exists, by the convexity of C, at least one $i_o \in \{1,\dots,9\}$ s.t. $\partial C \cap (I_{i_o}^1)^o = \emptyset$, and therefore $\nu(\partial C) \leqq \frac{8}{9}\nu(K_2)$. In the second step consider $I_{i_1}^1$ with $\partial C \cap (I_{i_1}^1)^o \neq \emptyset$ and choose again a subdivision of $I_{i_1}^1$ into nine subintervals of the form $I_{i_1 i}^2 = I_{j_i}^2 \times I_{k_i}^2$ where $I_{j_i}^2$ and $I_{k_i}^2$ are such that $\mu_1(I_{j_i}^2) = \frac{1}{3}\mu_1(I_{j_{i_1}}^1)$ and $\mu_2(I_{k_i}^2) = \frac{1}{3}\mu_2(I_{k_{i_1}}^1)$; then $\nu(I_{i_1 i}^2) = \frac{\nu(K_2)}{9^2}$, $i=1,\dots,9$. By the convexity of C it follows as before that $\partial C \cap (I_{i_1 i_o}^2)^o = \emptyset$ for at least one $i_o \in \{1,\dots,9\}$ and therefore (noticing that there are at most eight I_i^1 with $\partial C \cap (I_i^1)^o \neq \emptyset$) it follows that $\nu(\partial C) \leqq 8 \cdot \frac{8}{9^2}\nu(K_2)$. By continuing this procedure we obtain in the n-th step

$\nu(\partial C) \leq (\frac{8}{9})^n \nu(K_2)$ and therefore (with $n\to\infty$) $\nu(\partial C)=0$. □

The proof of the above lemma in this form essentially goes back to P.E. Hansen. For a different proof cf. [13] , Lemma 3.4.

3. Glivenko-Cantelli theorems.

3.1 Theorem. Let $\mathcal{C} \subset \mathcal{B}_k$, $k\geq 1$, be a $(\mu,\mathcal{A})$-uniformity class and $(\xi_i)_{i\in\mathbb{N}}$ be a sequence of i.i.d. random vectors defined on some p-space $(\Omega,\mathcal{F},\mathbb{P})$ with values in $\mathbb{R}^k$ and p.d. $\mu|\mathcal{B}_k$. Then there exists $\Omega_o \in \mathcal{F}$ with $\mathbb{P}(\Omega_o)=1$ such that $\lim_{n\to\infty} (\sup_{C\in\mathcal{C}} |\mu_n^\omega(C)-\mu(C)|)=0$ for all $\omega\in\Omega_o$.

Proof. By 2.4 there exists a countable sub-field $\mathcal{A}_o$ of $\mathcal{A}$ s.t. $\mathcal{C}$ is also a $(\mu,\mathcal{A}_o)$-uniformity class. It follows from the strong law of large numbers that there exists $\Omega_o \in \mathcal{F}$ with $\mathbb{P}(\Omega_o)=1$ s.t. $\lim_{n\to\infty} \mu_n^\omega(A)=\mu(A)$ for all $A \in \mathcal{A}_o$. The assertion now follows from the uniformity of $\mathcal{C}$. □

As a simple consequence the preceding theorem together with 2.9 and 2.13, yield the following Glivenko-Cantelli results.

3.2 Theorem (Glivenko-Cantelli). Let $(\xi_i)_{i\in\mathbb{N}}$ be a sequence of i.i.d. random vectors defined on some p-space $(\Omega,\mathcal{F},\mathbb{P})$ with values in $\mathbb{R}^k$, $k\geq 1$, and d.f. F; then there exists $\Omega_o \in \mathcal{F}$ with $\mathbb{P}(\Omega_o)=1$ such that $\lim_{n\to\infty} (\sup_{\underline{t}\in\mathbb{R}^k} |F_n^\omega(\underline{t})-F(\underline{t})|)=0$ for all $\omega\in\Omega_o$.

3.3 Corollary (Varadarajan). Under the assumption of 3.2 we have $\mu_n^\omega \xrightarrow[w]{} \mu$ for all $\omega\in\Omega_o$.

3.4 Theorem (R.R. Rao). Under the assumption of 3.2 it follows that $\lim_{n\to\infty} (\sup_{C\in\mathcal{C}_k} |\mu_n^\omega(C)-\mu(C)|)=0$ for all $\omega\in\Omega_o$, whenever μ fulfills (+) $\sup_{C\in\mathcal{C}_k} \mu_c(\partial C) = 0$.

Acknowledgement. The authors are especially grateful to F. Topsøe for the useful discussions they had during a recent stay at Copenhagen which was partly supported by the Danish Natural Science Research Council.

REFERENCES

[1] Cantelli, F.P. (1933). Sulla determinazione empirica delle leggi die probabilita. Ist. Ital. Attnari 4, 421-424.

[2] Dehardt, J. (1971). Generalizations of the Glivenko-Cantelli theorem. Ann. Math. Statist. 42, 2050-2055.

[3] Elker, J. (1975). Über ein gleichmäßiges Gesetz der großen Zahlen. Diplomarbeit. Ruhr-Universität Bochum.

[4] Fabian, V. (1970). On uniform convergence of measures. Z. Wahrscheinlichkeits-theorie verw. Gebiete 15, 139-143.

[5] Gaenssler, P. (1973). On convergence of sample distributions. Bull. Internat. Statist. Inst. 45,1, 427-432.

[6] Gaenssler, P. (1974). Around the Glivenko-Cantelli theorem. In: Limit theorems of Probability Theory; Keszthely. Ed. by P. Révész, 93-103.

[7] Gaenssler, P. and Stute, W. (1975/76). A survey on some results for empirical processes in the i.i.d. case. RUB-Preprint Series No. 15.

[8] Glivenko, V. (1933). Sulla determinazione empirica della legge die probabilita. Giorn. Ist. Ital. Attnari 4, 92-99.

[9] Halmos, P. (1958). Measure Theory. Princeton: Van Nostrand.

[10] Kiefer, J. (1961). On large deviations of the empiric D.F. of vector chance variables and a law of iterated logarithm. Pacif. J. Math. 11, 649-660.

[11] Krickeberg, K. (1976). An alternative approach to Glivenko-Cantelli theorems. In this volume.

[12] Rao, R.R. (1962). Relations between weak and uniform convergence of measures with appl.. Ann. Math. Statist. 33, 659-680.

[13] Stute, W. (1976). On a generalization of the Glivenko-Cantelli theorem. Z. Wahr-scheinlichkeitstheorie verw. Gebiete 35, 167-175.

[14] Stute, W. (1974). On uniformity classes of functions with an application to the speed of mean Glivenko-Cantelli convergence. RUB-Preprint Series No. 5.

[15] Stute, W. (1975). Convergence rates for the isotrope discrepancy. To appear in Ann. Probability.

[16] Topsøe, F. (1970). Topology and Measure. Lecture Notes in Mathematics 133. Berlin: Springer.

[17] Topsøe, F., Dudley, R.M. and Hoffmann-Jørgensen, J. (1976). Two examples concerning uniform convergence of measures w.r.t. balls in Banach spaces. In this volume.

[18] Topsøe, F. (1976). Uniformity in convergence of measures. Preprint Series No. 27. University of Copenhagen.

[19] Valentine, F. (1968). Konvexe Mengen. Mannheim: Bibliographisches Institut. Band 402/402 a.

[20] Wolfowitz, J. (1954). Generalizations of the theorem of Glivenko-Cantelli. Ann. Math. Statist. 25, 131-138.

[21] Wolfowitz, J. (1960). Convergence of the empiric distribution function on half-spaces. Contrib. to Probability and Statist., 504-507.

AN ALTERNATIVE APPROACH TO GLIVENKO-CANTELLI THEOREMS

Klaus Krickeberg
Université Paris V
12 rue Cujas
F-75005 Paris

Introduction

The present note was not read as a formal lecture at the meeting in Oberwolfach, but was the subject of some discussions. Its purpose is mainly methodical, namely to outline an alternative method to derive Glivenko-Cantelli theorems in several dimensions. Like Gänssler in [1] we use the uniform law of large numbers given there (see the Theorem below), but we apply it somewhat more directly by introducing an appropriate topology into the class of the subsets in question of the given basic space. In this way we also avoid the use of uniform classes (see Gänssler and Stute [2]). Finally we discuss some extensions of the range of applicability of the method.

The uniform law of large numbers

We consider throughout a basic probability space $(V, \mathcal{V}, \mu)$ and a sequence of independent random elements $\xi_1, \xi_2, \ldots$ which are identically distributed in V according to the law μ and defined on some probability space $(\Omega, \mathcal{F}, P)$.

Theorem (Gänssler). Let T be a compact space and $(h_t)_{t \in T}$ a family of extended real-valued functions defined on V which has the following properties:

1. Every h_t is $\mathcal{V}$-measurable.
2. There is a basis $\mathcal{T}$ of the topology of T which consists of open sets such that:

   2i. For every $S \in \mathcal{T}$ the functions $\inf\{h_t : t \in S\}$ and $\sup\{h_t : t \in S\}$ are $\mathcal{V}$-measurable and μ-integrable.

   2ii. For every $t_o \in T$ and $\delta > 0$ there exists a neighbourhood $S \in \mathcal{T}$ of t_o such that

$$\mu(h_{t_o}) - \varepsilon < \mu(\inf h_s : s \in S\}) \qquad (1)$$
$$\leq \mu(\sup h_s : s \in S\}) < \mu(h_{t_o}) + \varepsilon .$$

Then for P-almost all ω:

$$\lim_{n\to\infty} \sup_{t\in T} \left| \frac{1}{n} \sum_{i=1}^{n} h_t(\xi_i(\omega)) - \mu(h_t) \right| = 0 . \qquad (2)$$

<u>Proof</u>. Given the situation of condition 2ii we have

$$\mu(h_{t_o}) - \varepsilon < \mu(h_t) < \mu(h_{t_o}) + \varepsilon$$

for all $t \in S$ which means that the function $t \mapsto \mu(h_t)$ is continuous. Therefore, given again t_o and ε, if S is selected as in 2ii, the set

$$\{t \in S: \mu(\inf\{h_s : s \in S\}) > \mu(h_t) - \varepsilon\} \qquad (3)$$

is open and contains t_o. Hence there exists a set $S(t_o) \in \mathcal{C}$ which includes t_o and is contained in the set (3). This implies that for every $t \in S(t_o)$:

$$\mu(\inf\{h_s : s \in S(t_o)\}) \geq \mu(\inf\{h_s : s \in S\}) > \mu(h_t) - \varepsilon$$

and thus

$$\frac{1}{n} \sum_{i=1}^{n} (h_t \circ \xi_i - \mu(h_t)) \geq \frac{1}{n} \sum_{i=1}^{n} \left[\inf\{h_s \circ \xi_i : s \in S(t_o)\} - \mu(\inf\{h_s : s \in S(t_o)\}) \right] - \varepsilon .$$

As T is compact, there are points $t_1, \ldots, t_m \in T$ such that $T = S(t_1) \cup \ldots \cup S(t_m)$. For every $t \in T$ we then have

$$\frac{1}{n} \sum_{i=1}^{n} (h_t \circ \xi_i - \mu(h_t)) \geq$$
$$\geq \min_{j=1,\ldots,m} \frac{1}{n} \sum_{i=1}^{n} \left[\inf\{h_s \circ \xi_i : s \in S(t_j)\} - \mu(\inf\{h_s : s \in S(t_j)\}) \right] - \varepsilon$$

for every n and therefore

$$\varliminf_{n\to\infty} \inf_{t\in T} \frac{1}{n} \sum_{i=1}^{n} \left[h_t \circ \xi_i - \mu(h_t) \right] \geq$$
$$\geq \min_{j=1,\ldots,m} \lim_{n\to\infty} \frac{1}{n} \sum_{i=1}^{n} \left[\inf\{h_s \circ \xi_i : s \in S(t_j)\} - \mu(\inf\{h_s : s \in S(t_j)\}) \right] - \varepsilon .$$

By applying the strong law of large numbers for $j = 1,\ldots,m$ to the function $\inf\{h_s : s \in S(t_j)\}$ in $L_1(\mu)$ we find sets $\Delta_j(\varepsilon) \in \mathcal{F}$ such that $P(\Delta_j(\varepsilon)) = 0$ and

$$\lim_{n\to\infty} \frac{1}{n} \sum_{i=1}^{n} \left[\inf\{h_s(\xi_i(\omega)):\ s\in S(t_j)\} - \mu(\inf\{h_s:\ s\in S(t_j)\})\right] = 0$$

for every $\omega \notin \Delta_j(\varepsilon)$, hence

$$\underline{\lim}_{n\to\infty} \inf_{t\in T} \frac{1}{n} \sum_{i=1}^{n} \left[h_t(\xi_i(\omega)) - \mu(h_t)\right] \geq -\varepsilon$$

for every $\omega \notin \Delta(\varepsilon) = \bigcup_{j=1}^{m} \Delta_j(\varepsilon)$. By setting $\Delta = \bigcup_{l=1}^{\infty} \Delta(1/l)$ we have $P(\Delta)=0$, and $\omega \notin \Delta$ implies

$$\underline{\lim}_{n\to\infty} \inf_{t\in T} \frac{1}{n} \sum_{i=1}^{n} \left[h_t(\xi_i(\omega)) - \mu(h_t)\right] \geq 0.$$

The opposite inequality for $\overline{\lim}$ is obtained in the same way for P-almost all ω which gives (2).

Glivenko-Cantelli theorems

In this chapter, T will always be a subset of $\mathcal{V}$, and $h_t = 1_t$ will be the indicator function of the set $t \in T$. The condition 1 of the Theorem is then automatically satisfied. If we can introduce into T a compact topology and if we can verify the condition 2, we obtain the Glivenko-Cantelli theorem for T and μ, that is for P-almost all ω:

$$\lim_{n\to\infty} \sup_{t\in T} |\mu_n^\omega(t) - \mu(t)| = 0 \tag{4}$$

where

$$t \mapsto \mu_n^\omega(t) = \frac{1}{n}\sum_{i=1}^{n} 1_t(\xi_i(\omega))$$

is the empirical distribution given by ω and n.

Let us note that with the present setup we have for every $S \subseteq T$:

$$\inf\{h_t:\ t\in S\} = 1_{\bigcap\{t:\ t\in S\}},$$
$$\sup\{h_t:\ t\in S\} = 1_{\bigcup\{t:\ t\in S\}}. \tag{5}$$

Example 1. Here, V=R will be the real line, $\mathcal{V}$ the sigma-algebra of its Borel sets, and T the class of the sets of the form $\emptyset$, R, $]-\infty,u[$ and $]-\infty,u]$ with $u\in R$. We use a topology which had been employed for different purposes before and which is defined in the following way: a basis of neighbourhoods of $t_0 = \emptyset$ or $t_0 =]-\infty,u]$ consists of all sets of the form $\{s\in T:\ t_0 \subseteq s \subseteq t\}$ where t runs through all sets in T such that $t_0 \subsetneq t$ and

$t_0 \neq t$. Analogously, a basis of neighbourhoods of the form $\{s \in T: t \subseteq s \subseteq t_0\}$ with $t \in T$, $t \subseteq t_0$ and $t \neq t_0$ is used when $t_0 = R$ or $t_0 =]-\infty, u[$.

Proposition 1. The above topology is compact and has no countable base.

We will give a short proof of these well known facts. For the first one it suffices to show that every family $(S_t)_{t\in T}$ where S_t is a neighbourhood of t of the form given above, contains a finite covering of T. In order to deduce this from the compactness of the extended real line $\overline{R} = R \cup \{-\infty, \infty\}$, consider the following map β from T onto $\overline{R}$:

$$\beta(\emptyset) = -\infty, \quad \beta(R) = +\infty, \quad \beta(]-\infty, u[) = \beta(]-\infty, u]) = u .$$

Define for every $u \in \overline{R}$ an open subset I_u of $\overline{R}$ which contains u by setting

$$I_{-\infty} = [-\infty, \beta(t)[\quad \text{if} \quad S_{\emptyset} = \{s \in T: s \subseteq t\};$$

$$I_{+\infty} =]\beta(t), +\infty] \quad \text{if} \quad S_R = \{s \in T: t \subseteq s\};$$

$$I_u =]\beta(t'), \beta(t'')[\quad \text{if } u \in R \text{ and}$$

$$S_{]-\infty, u[} = \{s \in T: t' \subseteq s \subseteq]-\infty, u[\}, \quad S_{]-\infty, u]} = \{s \in T:]-\infty, u] \subseteq s \subseteq t''\}.$$

There is a finite covering $I_{u_1}, \ldots, I_{u_l}$ of $\overline{R}$, and this implies

$$T \subseteq \bigcup_{j=1}^{l} (S_{]-\infty, u_j[} \cup S_{]-\infty, u_j]})$$

where $S_{]-\infty, u_j[}$ or $S_{]-\infty, u_j]}$ is, of course, absent whenever $u_j = -\infty$ or $u_j = +\infty$, respectively.

Next, let $\mathcal{C}$ be any base of the topology of T. Then for every $u \in R$ select a neighbourhood $S_u \in \mathcal{C}$ of the element $]-\infty, u] \in T$. Since

$$u = \inf\{v:]-\infty, v] \in S_u\},$$

the set S_u determines u, that is, the map $u \mapsto S_u$ is injective, and hence $\mathcal{C}$ is not countable.

The condition 2 of the Theorem is now trivially satisfied. In fact, consider any neighbourhood $S = \{s \in T: t_0 \subseteq s \subseteq t\}$ of $t_0 =]-\infty, u_0]$. By (5),

$$\inf\{h_s: s \in S\} = 1_{]-\infty, u_0]} = h_{t_0} ,$$

$$\sup\{h_s: s \in S\} = 1_{]-\infty, u[} \quad \text{where } u = \beta(t),$$

and

$$\lim_{u \searrow u_0} \mu(]-\infty, u[) = \mu(]-\infty, u_0]).$$

The same reasoning applies when $t_0 =]-\infty, u_0[$ or $t_0 = \emptyset$ or $t_0 = R$. As a result we get from (4) the classical Glivenko-Cantelli theorem on empirical cumulative distribution functions.

<u>Example 2</u>. We take again $V=R$, $\mathfrak{V}$ the Borel sets, and T the class of all connected subsets of V including the empty set $\emptyset$. The formula (4) follows then immediately from that proved in example 1, but we can also introduce into T a compact topology so as to satisfy the condition 2 of the Theorem. For example a typical neighbourhood of an interval $t_0 =]u_0, v_0]$ with $-\infty < u_0 < v_0 < \infty$ consists of all sets t of the following type:

$$t =]u,v] \quad \text{with } u_0 \leq u \underset{(-)}{<} u', \quad v_0 \leq v \underset{(-)}{<} v',$$

$$t =]u,v[\quad \text{with } u_0 \leq u \underset{(-)}{<} u', \quad v_0 < v \underset{(-)}{<} v',$$

$$t = [u,v] \quad \text{with } u_0 < u \underset{(-)}{<} u', \quad v_0 \leq v \underset{(-)}{<} v',$$

$$t = [u,v[\quad \text{with } u_0 < u \underset{(-)}{<} u', \quad v_0 < v \underset{(-)}{<} v',$$

where u' and v' are fixed, $u_0 < u'$, $v_0 < v'$, and $\underset{(-)}{<}$ at any given place is either to be read throughout as $\leq$ or throughout as $<$. A basis of neighbourhoods of $t_0 = \{u_0\}$ where $u_0 \in R$ is made up of the sets of the type $S = \{s \in T:\ u_0 \in s \subseteq t\}$ where t may be any interval which contains u_0 in its interior. Neighbourhoods of half-lines and of the entire line R are defined analogously.

The definition of neighbourhoods of the empty set is slightly more complicated. Consider a map q which assigns to every $u \in \overline{R}$ a set $q(u)$ of the following form:

if $u \in R$, then $q(u) = t \setminus \{u\}$ where $t \in T$ and $u \in \text{interior}(t)$;

if $u = \pm\infty$, then $q(u) = t \cup \{u\}$ where $t \in T$ and $u \in \text{interior}(t \cup \{u\})$.

Set $S(q,u) = \{s:\ s \in T,\ s \subseteq q(u)\}$ and $S(q) = \bigcup \{S(q,u):\ u \in \overline{R}\}$. Then the class of all $S(q)$ where q runs through all these maps, is a basis of neighbourhoods of $\emptyset$. Intuitively speaking, there is one "copy" $\emptyset_u$ of the empty set for every $u \in \overline{R}$ together with its obvious neighbourhoods, and we then lump all $\emptyset_u$ together to form the single "point" $\emptyset$ of T. For example, the sequence $[\frac{1}{n+1}, \frac{1}{n}]$ as well as the sequence $[n, n+1]$ converges in T to $\emptyset$ whereas $[-\frac{1}{n}, \frac{1}{n}]$ and $[0, \frac{1}{n}]$ converge to $\{0\}$.

The compactness of this topology including the fact that it is separated can now be proved as in example 1.

<u>Example 3</u>. Here, $V=R^k$, $\mathfrak{V}$ is the class of the Borel sets of V and T that of all generalized rectangles

$$t = \mathop{\times}_{i=1}^{k} t_i \tag{6}$$

where each t_i may be any of the sets of example 2. A neighbourhood of t consists of all sets of the form

$$s = \bigtimes_{i=1}^{k} s_i$$

where $s_1,\dots,s_k$ run through given neighbourhoods of $t_1,\dots,t_k$, respectively. Essentially, T is the k-fold cartesian product of the space of example 2 except that the empty set can be represented in the form (6) in many ways, and hence we have again a lumping of elements into a single element $\emptyset$ of T. This and the compactness of the product of compact spaces gives the compactness of T. The condition 2 of the Theorem is easily verified, hence (4) holds again for any μ. By restricting (6) to products of sets from example 1, that is, to generalized quadrants or octants, we obtain the Glivenko-Cantelli theorem for empirical multidimensional cumulative distribution functions.

Example 4. Let $\mathcal{C}$ be the class of all non-empty closed convex subsets of $V=R^k$. For every $u\in V$ and $A\subseteq V$, $A\neq\emptyset$, set

$$\delta(u,A)=\inf\{|u-v| : v\in A\}.$$

The function $\delta(\cdot,A)$ being continuous, the sets

$$A^{\bar{\eta}}=\{u\in V: \delta(u,A)\le\eta\}\ ,\qquad A^{\underline{\eta}}=\{u\in\bar{A}: \delta(x,\partial A)\ge\eta\}$$

where $\bar{A}$ denotes the closure and ∂A the boundary of A, are closed for every $\eta\ge 0$. Recall that for any two non-empty subsets A and B of V the number

$$\begin{aligned} d(A,B) &= \max\{\sup\{\delta(u,B): u\in A\},\ \sup\{\delta(v,A): v\in B\}\} \\ &= \inf\{\eta\in R: A\subseteq B^{\bar{\eta}},\ B\subseteq A^{\bar{\eta}}\} \end{aligned}$$

is called the Hausdorff distance of A and B, and that the following is true:

Proposition 2. For every compact set $K\in\mathcal{C}$, the set $\mathcal{C}_K=\{C: C\in\mathcal{C},\ C\subseteq K\}$ is compact in the topology induced by d.

For a proof, see e.g. Matheron [3].

Proposition 3 (R.R.Rao). Let μ_0 be the diffuse (non-atomic) part of μ and suppose that $\mu_0(\partial C)=0$ for every $C\in\mathcal{C}$. Then μ satisfies the Glivenko-Cantelli theorem with $\mathcal{C}$, that is for P-almost all ω:

$$\lim_{n\to\infty}\ \sup_{C\in\mathcal{C}} |\mu_n^{\omega}(C)-\mu(C)| = 0. \tag{7}$$

Proof. Assume first that μ is diffuse, that is $\mu=\mu_0$. Let K_r be the closed ball of radius r around the origin. By the usual reasoning we see that it suffices to prove (7) for every $r>0$ and P-almost all ω with $\mathcal{C}_{K_r}$ instead of $\mathcal{C}$. We will apply the Theorem with $T=\mathcal{C}_{K_r}$ where $\mathcal{T}$ is the set of all open balls in T for d.

Let S be such a ball around $t_0\in T$:

$$S = \{s\in T:\ d(s,t_0)<\eta\} \tag{8}$$

where $\eta>0$. We are going to prove that

$$\bigcap_{s\in S} s = \{u\in t_0:\ \delta(u,\partial t_0)\geq\eta\} = t_o^{\eta}, \tag{9}$$

$$\bigcup_{s\in S} s = \{u\in K_r:\ \delta(u,t_0)<\eta\} = K_r\cap \mathrm{int}(t_0^{\bar\eta}) . \tag{10}$$

We note first that for $\eta'>0$:

$$d(t_0,t_0^{\eta'})=\eta' \quad \text{if } t_0^{\eta'}\neq\emptyset , \tag{11}$$

$$d(t_0,\ K_r\cap t_0^{\bar\eta'})\leq\eta'. \tag{12}$$

Moreover, using elementary geometric properties of convex sets we can show that

$$u\in t_0,\ s\in\mathcal{C}_{K_r},\ d(s,t_0)<\eta\ ,\ \delta(u,\partial t_0)\geq\eta \Longrightarrow u\in s . \tag{13}$$

To prove (9), suppose that $u\in\bigcap\{s:\ s\in S\}$. For $0<\eta'<\eta$ we have $t_0^{\eta'}\in S$ by (8) and (11), hence $u\in t_0^{\eta'}$, that is $\delta(u,\partial t_0)\geq\eta'$. Since this holds for every $\eta'<\eta$ we obtain $u\in t_o^{\eta}$. Conversely, suppose that this is true, that is $u\in t_0$ and $\delta(u,\partial t_0)\geq\eta$. Then if $s\in S$, we have $d(s,t_0)<\eta$, and $u\in s$ now follows from (13).

To derive (10), consider a point $u\in\bigcup\{s:\ s\in S\}$. Let $s\in S$ be such that $u\in s$. Then $d(s,t_0)<\eta$ implies $\delta(u,t_0)<\eta$. Conversely, suppose that $\delta(u,t_0)<\eta$ where $u\in K_r$. Select η' such that $\delta(u,t_0)<\eta'<\eta$ and set $s=K_r\cap t_0^{\bar\eta'}$. By (12), $d(s,t_0)\leq\eta'<\eta$, hence $s\in S$ and therefore $u\in\bigcup\{s:\ s\in S\}$.

The condition 2i of the Theorem, with $h_t=1_t$, is now trivally satisfied on account of (5), (9) and (10). Moreover, if $\mu(\partial t_0)=0$, we have

$$\mu(t_0)=\lim_{\eta\to0}\mu(t_o^{\eta})=\lim_{\eta\to0}\mu(K_r\cap\mathrm{int}(t_0^{\bar\eta}))$$

which proves that the condition 2ii also holds.

In the case $\mu_o=0$, that is the case of a discrete law μ, the relation (7) is almost trivial; see [2]. The general case $0<\mu_0(V)<1$ can be reduced to the two particular ones $\mu_0(V)=1$ and $\mu_0(V)=0$ by the following more or less standard procedure which we will outline.

As (7) depends only on the law of the sequence $\xi_1, \xi_2,\ldots$, we may construct this sequence as convenient. Set $p=\mu_0(V)$ and $q=1-p$, and let (ξ_i') and (ξ_i'') be independent sequences of independent random elements distributed in V according to the law μ_0/p and $(\mu-\mu_0)/q$, respectively.

Let (ϑ_i) be a sequence of independent random variables with $P\{\vartheta_i=1\}=p$ and $P\{\vartheta_i=2\}=q$ which is independent of the two sequences (ξ_i') and (ξ_i''). Define ξ_i by

$$\xi_i = \begin{cases} \xi_i' & \text{if } \vartheta_i = 1, \\ \xi_i'' & \text{if } \vartheta_i = 2, \end{cases}$$

and set

$$n'(\omega) = \#\{i:\ 1 \le i \le n,\ \vartheta_i = 1\}, \qquad n''(\omega) = \#\{i:\ 1 \le i \le n,\ \vartheta_i = 2\}.$$

By the law of large numbers we have for almost all ω that $n'(\omega)/pn \longrightarrow 1$ and $n''(\omega)/qn \longrightarrow 1$ and this, together with (7) applied to μ_0/p and (ξ_i') as well as to $(\mu-\mu_0)/q$ and (ξ_i'') implies (7) for μ and (ξ_i).

Extensions

The first extension concerns the class T of sets. We note that each of the topologies employed in the examples 1-3 is nothing but the topology of pointwise convergence of the indicator functions of the sets in question. In the space $\{0,1\}^V$ of the indicator functions of all subsets of V, this topology is the product of the discrete topology in the factor spaces $\{0,1\}$, and it makes $\{0,1\}^V$ compact by Tihonov's theorem. If we consider as before the sets themselves instead of their indicators, or in other words the power set $\mathcal{P}(V)$ instead of $\{0,1\}^V$, then the various classes of sets which we have studied in the examples 1-3 are endowed with the relative topology of the pointwise, or product, topology. The proof of the compactness of any of these subsets T of $\mathcal{P}(V)$ amounts to the proof that they are closed in $\mathcal{P}(V)$.

If $V=R^k$, the class $\mathcal{V}$ of all Borel sets is dense in $\mathcal{P}(V)$ for the pointwise topology, and therefore not closed, but it is not hard to find other interesting subclasses T of $\mathcal{V}$ which are closed, and which moreover satisfy the condition 2 of the Theorem for any μ. We give some examples where the proof that T is closed in $\mathcal{P}(V)$ for the pointwise topology is fairly easy. We also remark at once that in these examples, as in the examples 1-3, every element of T has a countable fundamental system of neighbourhoods for the relative pointwise topology. The proof of the condition 2 is then essentially an application of Lebesgue's dominated convergence theorem.

Example 5. We are going to enlarge the class studied in example 3. Let $m \ge 2$ be a fixed positive integer. We define a class T_k of Borel subsets of $V=R^k$ by induction on k. For T_1 we take the class of all connected subsets of R (example 2). Having defined T_{k-1} we let T_k be the class of all convex, bounded or unbounded, polyhedra t in V of one of the following types.

i) $t=\emptyset$.

ii) The minimal affine subspace of V containing t has dimension $l<k$, and t belongs in this subspace to T_l.

iii) The interior of t is a non-empty intersection of at most m open half-spaces in V, and the intersection of t with any of its closed l-dimensional faces, for $l=1,\dots,k$, is of type ii).

From this example we obtain (4) in particular for the class of all closed convex sets which are intersections of at most m closed half-spaces. It should be noted that T may be larger than the closure in $\mathcal{P}(V)$ of the set of all intersections of at most m closed half-spaces; this happens already in the case k=m=2. The description of this closure in the general case would be fairly complicated.

In the following example, however, we will use such a non-constructive definition for a change.

<u>Example 6</u>. Let T be the closure in $\mathcal{P}(V)$, for the pointwise topology, of the set of all closed ellipsoids in $V=R^k$.

The explicit description of T is left to the reader as well as the construction of other examples, starting from families of convex bodies indexed by a finite number of real parameters.

The second extension bears on the "almost sure law" from which we started. This law can be regarded as a particular convergence theorem on random measures, because we can write

$$\frac{1}{n}\sum_{i=1}^{n} h(\xi_i(\omega)) = \mu_n^\omega(h)$$

for any real-valued $\mathcal{V}$-measurable function h, and $(\omega,h)\longmapsto\mu_n^\omega(h)$ is, for fixed n, the random measure in V which charges the points $\xi_1(\omega),\dots,\xi_n(\omega)$ with the mass 1/n each. An analysis of the proof of the Theorem shows that other almost sure convergence theorems on random measures also have uniform versions. We give two examples.

<u>A uniform ergodic theorem for random measures</u>. Let $V=R^k$ with the Borel sigma-algebra $\mathcal{V}$ and $\mathcal{H}$ the space of all $\mathcal{V}$-measurable and bounded functions with a compact carrier. A random measure ζ in V is a function $(\omega,h)\longmapsto\zeta(\omega,h)$ such that:

$h\longmapsto\zeta(\omega,h)$ is a positive Radon measure in V for every $\omega\in\Omega$;

$\omega \longmapsto \zeta(\omega, h)$ is $\mathcal{F}$-measurable for every $h \in \mathcal{H}$.

Denote by ϑ_u the translation operator $\vartheta_u v = v-u$ for $u, v \in V$, and by λ the Lebesgue measure in V. Assume that ζ has a finite intensity measure

$$\nu(h) = \int_{\Omega} \zeta(\omega, h) P(d\omega)$$

and is stationary under translations so that $\nu = z\lambda$ with a constant $z \geq 0$. Then the Palm distribution of ζ is well defined [4] as a certain random measure in V; let ν_0 be its intensity measure. For every $K \in \mathcal{V}$ with $0 < \lambda(K) < \infty$ set

$$\mu_K^{\omega}(h) = \frac{1}{\lambda(K)} \int_K \zeta(\omega, h \circ \vartheta_u) \zeta(\omega, du)$$

so that $(\omega, h) \longmapsto \mu_K^{\omega}(h)$ is again a random measure in V.

Write K_r for the closed ball around the origin with radius r. A system $\mathcal{K}$ of closed convex subsets of V is called regular if $\sup\{\lambda(K): K \in \mathcal{K}\} = +\infty$ and there exist numbers $a > 0$, $r_0 > 0$ and a function $c: \mathcal{K} \to]0, +\infty[$ such that $K \subseteq K_{c(K)}$ for all $K \in \mathcal{K}$ and $\lambda(K) \geq a\,\lambda(K_{c(K)})$ for all $K \in \mathcal{K}$ with $c(K) \geq r_0$. The ergodic theorem for random measures of Nguyen Xuan Xanh and H. Zessin [5] states that for P-almost all ω:

$$\lim_{\lambda(K) \to \infty} \mu_K^{\omega}(h) = z\nu_0(h)$$

whenever $\nu_0(|h|)$ is finite. The uniform version looks like this:

Let $(h_t)_{t \in T}$ be a family of functions which satisfies the assumptions of the Theorem with $\mu = \nu_0$. Then for P-almost all ω we have

$$\lim_{\lambda(K) \to \infty} \sup_{t \in T} \left| \mu_K^{\omega}(h_t) - z\nu_0(h_t) \right| = 0.$$

Examples for such families $(h_t)_{t \in T}$ can be formed as in the preceding chapter.

Estimation of the intensity of stationary line processes. Let V be S_1, the unit circle of the plane R^2, and μ a positive Radon measure on V. The problem is to estimate μ from the observation of a realization of a suitable Poisson line process in R^2. As usual we represent an oriented line x in R^2 by the angle φ which it makes with a fixed line through the origin, and its signed distance p form the origin, p being positive, for example, if the origin lies on the left bank of x. In this way the set of all oriented lines becomes the product space $X = V \times R$. A line process in R^2 can thus be regarded as a point process on X. The most general Poisson line process which is stationary under translations of R^2 is the Poisson process on X with an intensity measure of the form $\mu \otimes \lambda$ where λ denotes the Lebesgue measure on R. It is this process which we are going to consider.

Let K_r be the closed disk in R^2 of radius r around the origin; r will be kept constant for a while. For any realization ω of the process, let $\varphi_1(\omega),\dots,\varphi_{n(\omega)}(\omega)$ be the angles of those lines of the realization which hit K_r. Set

$$\mu_r^\omega(h) = \frac{1}{2r}\sum_{i=1}^{n(\omega)} h(\varphi_i(\omega)).$$

Then $(\omega,h)\longmapsto\mu_r^\omega(h)$ is a random measure on V. For fixed μ-integrable h we have for P-almost all ω:

$$\lim_{r\to\infty}\mu_r^\omega(h) = \mu(h)$$

(see [6]), and we can then show that in fact P-almost surely

$$\lim_{r\to\infty}\ \sup_{t\in T}\left|\mu_r^\omega(h_t)-\mu(h_t)\right| = 0$$

if $(h_t)_{t\in T}$ satisfies the conditions of the Theorem. In particular, this is true for the class of the indicator functions of all segments of S_1.

The generalization to the case of flat processes in R^k of a given dimension l with $1\leq l\leq k$ is fairly immediate. Here V would be the compact manifold of all l-dimensional linear subspaces of R^k.

Bibliography

1. Gänssler, P.: Around the Glivenko-Cantelli theorem. Colloquia Math. Societatis János Bolyai 11, 93-103 (1974)
2. Gänssler, P. and Stute, W.: On uniform convergence of measures with applications to uniform convergence of empirical distributions. In this volume.
3. Matheron, G.: Random sets and integral geometry. New York: Wiley 1975
4. Kerstan, J., Matthes, K. and Mecke, J.: Unbegrenzt teilbare Punktprozesse. Berlin: Akademie-Verlag 1974
5. Nguyen Xuan Xanh and Zessin, H.: Punktprozesse mit Wechselwirkung. To appear in Z. Wahrscheinlichkeitstheorie verw. Gebiete.
6. Fellous, A. and Granara, J.: Thèses 3e cycle. Université René Descartes, Paris 1976

WEAK CONVERGENCE UNDER CONTIGUOUS ALTERNATIVES OF THE EMPIRICAL PROCESS WHEN PARAMETERS ARE ESTIMATED: THE D_k APPROACH

G. Neuhaus
Math. Institute, University of Giessen, W.-Germany

1. INTRODUCTION

Current papers dealing with asymptotic statistics are usually based on the concept of weak convergence, and a great part of them are concerned with just the same sort of problems as in Doob's [7] foundation laying "heuristic approach" paper, namely by showing that certain test statistics are convergent in distribution (= weakly convergent) under the null hypothesis H_o. Such a result is needed to bound the asymptotic size of the corresponding test by a prescribed level of significance.

In order to learn something about the asymptotic performance of the test one needs parallel weak convergence results under alternatives, too. For the latter problems the additional concept of contiguity (see sec.2) has turned out to be very useful. It goes back to Le Cam [12], and has been made popular in non-parametric statistics by the book of Hâjek and Šidák [9].

A pecularity of this book is that the use of contiguity is limited to cases where the so-called Le Cam's third lemma is applicable, and this entails that only restricted alternatives like translation - or scale-alternatives can be treated. But from the papers of Behnen [1], [2], who mainly was concerned with tests based on linear rank statistics, it has become clear that the asymptotic power can be calculated for <u>all</u> contiguous alternatives by a method which makes in his case Le Cam's third lemma dispensable. The essentials of that method are contained in Behnen and Neuhaus [3].

Looking at the asymptotic power Behnen [2] showed that it might be dangerous to rely on results of asymptotic power for special contiguous alternatives like translation-alternatives, because for other contiguous alternatives which even may look like translation-alternatives the asymptotic power may be quite different, perhaps much worse than expected. Therefore, it seems of more value to consider in asymptotic power studies of tests under local alternatives the whole class of contiguous alternatives instead of special subclasses.

The present paper shows that for tests based on the empirical distribution function (df), especially Kolmogorov-Smirnov (KS)- and Cramér-von Mises (CvM)-tests, the concept of weak convergence (of the empirical process) and the concept of contiguity fit very well.

To become specific, let $U_1^n,\dots,U_n^n, n\geq 1$, be a triangular array of rowwise i.i.d. random k-vectors each having a continuous df. Under the composite <u>null-hypothesis</u> H_o, for all n these df's are equal and belong to some given class $\mathcal{R}=\{F_\theta,\theta\in\Theta\}$ of df's $F_\theta=F(\cdot,\theta)$ on $\mathbb{R}_k, k\geq 1$. It is not assumed that $\mathcal{R}$ is dominated. The parameter space Θ is an open subset of $\mathbb{R}_r, r\geq 1$, and under H_o the unknown $\theta=(\theta_1,\dots,\theta_r)'$ (the prime denotes the transposed vector) is estimated by a certain sequence of estimators $\hat\theta_n=\hat\theta_n(U_1^n,\dots,U_n^n), n\geq 1$.

It is an old idea (see e.g. Cramér [5]) to base a test for H_o on the (modified) KS-statistic

$$K_n=K_n(U_1^n,\dots,U_n^n) = \sup_{s\in\mathbb{R}_k} |n^{1/2}(F_n(s)-F(s,\hat\theta_n))|$$

or on the (modified) CvM-statistic

$$C_n=C_n(U_1^n,\dots,U_n^n) = \int [n^{1/2}(F_n(s)-F(s,\hat\theta_n))]^2 F(ds,\hat\theta_n),$$

where F_n is the empirical df of $U_1^n,\dots,U_n^n$. The first who studied weak convergence (always for $n\to\infty$) of C_n under H_o in the special case k=1 (one-dimensional rv's) and r=1 (one-dimensional parameters) was Darling [6], and recently Sukhatme [19] extended his results to the case r>1. Both authors proceed in the spirit of Doob's "heuristic approach" by showing weak convergence of the finite dimensional distributions (f.d.d.'s) of the process $Z_n=n^{1/2}(F_n-F(\cdot,\hat{\theta}_n))$ (more exactly of a reduced version $\bar{Z}_n$, see (1.5)) to those of a Gaussian limiting process $\bar{Z}$ on [0,1] with continuous paths. In some special cases weak convergence of the f.d.d.'s entails weak convergence of C_n, see e.g. Kac, Kiefer and Wolfowitz [11] and Kac [10], but in general this is not sufficient. While Sukhatme [19] cites Donsker's Theorem to ensure weak convergence $\bar{Z}_n \xrightarrow{\mathcal{D}} \bar{Z}$, which seems to be not directly applicable, Darling [6] reduces the problem after an "auxiliary assumption" to the case where θ is known, but the reduction seems to be not correct. In a recent paper Durbin [8] presented for the case $k=1, r\geq 1$ a complete proof for the weak convergence of a certain process $\tilde{Z}_n$ (being asymptotically equivalent to $\bar{Z}_n$, see Remark 2.3) not only under H_0 but also under certain alternatives. (As a matter of fact the assumption (1.3) (below) has to be added in Durbin's paper to make his proofs correct.) Durbin's alternatives are more or less parametric in nature and in practically all examples they are contiguous to H_0.

Here we shall show weak convergence of $\bar{Z}_n$ (or $\tilde{Z}_n$) for the case $k\geq 1, r\geq 1$ under the whole class of contiguous alternatives. In two earlier papers Neuhaus [13], [14] the present author started directly with the process $Z_n=n^{1/2}(F_n-F(\cdot,\hat{\theta}_n))$ regarding it as a rv with values in the Hilbert space $L_2(\mathbb{R}_k, F_\theta)$. The "Hilbert space approach" fits very well to CvM-statistics, whereas for KS-statistics the "function space approach" of this paper is more adequate.

Let us now summarize the necessary assumptions and regularity conditions: The sequence of estimators $\hat{\theta}_n=\hat{\theta}_n(U_1^n,\ldots,U_n^n), n\geq 1$, can be written as

(1.1) $n^{1/2}(\hat{\theta}_n-\theta) = n^{-1/2}\sum_{j=1}^{n} h(U_j^n,\theta) + R_n$,

$R_n=R_n(U_1^n,\ldots,U_n^n)=o_{F_\theta}(1)$, for some r-vectors $h(\cdot,\theta)=(h_1(\cdot,\theta),\ldots,h_r(\cdot,\theta))'$ fulfilling

(1.2) $\int h_i(\cdot,\theta)dF_\theta=0$, $b_{ij}(\theta)=\int h_i(\cdot,\theta)h_j(\cdot,\theta)dF_\theta$ finite, $\forall i,j=1,\ldots,r$, $\forall\,\theta\in\Theta$. $B(\theta)\equiv(b_{ij}(\theta))_{i,j=1,\ldots,r}$.

For the formulation of the regularity conditions for $\mathcal{P}=\{F_\theta,\theta\in\Theta\}$ write $\psi_\theta(s)=(F_1(s_1,\theta),\ldots,F_k(s_k,\theta))$, $s=(s_1,\ldots,s_k)\in\bar{\mathbb{R}}_k$, and $\bar{\psi}_\theta(t)=(F_1^{-1}(t_1,\theta),\ldots,F_k^{-1}(t_k,\theta))$, $t=(t_1,\ldots,t_k)\in E_k=[0,1]^k$, where $F_i(\cdot,\theta)$ is the i-th marginal of $F(\cdot,\theta)$ with left-continuous inverse $F_i^{-1}(\cdot,\theta)$, $i=1,\ldots,k$.

(1.3) For each $i=1,\ldots,k$ the intervals where $F_i(\cdot,\theta)$ is non-increasing remain unaltered when θ varies in Θ.

(1.4) The derivatives $\partial F(s,\theta)/\partial\theta_i$ exist $\forall s\in\bar{\mathbb{R}}_k$, $\forall\theta\in\Theta$, and the functions $q_i(t,\theta)\equiv\partial F(s,\theta)/\partial\theta_i$, $s=\bar{\psi}_\theta(t)$, are continuous on $E_k\times\Theta$, $i=1,\ldots,r$. $q(t,\theta)\equiv(q_1(t,\theta),\ldots,q_r(t,\theta))'$.

Now, let us assume that H_0 holds true with θ the underlying parameter, θ fixed in Θ. Then write $V_j^n=\psi_\theta(U_j^n)$, $j=1,\ldots,n$, and $\bar{F}(\cdot,\theta_1)=F(\bar{\psi}_\theta,\theta_1)\,\forall\theta_1\in\Theta$. Because of (1.3) the set $\{s\in\mathbb{R}_k:\bar{\psi}_\theta\circ\psi_\theta(s)=s\}$ is independent of $\theta\in\Theta$ and has $F(\cdot,\theta_1)$-probability one $\forall\theta_1\in\Theta$. Therefore, $\bar{F}(\cdot,\theta_1)$ is the df of V_j^n if U_j^n has df $F(\cdot,\theta_1)$. Denote by $\bar{F}_n$ the empirical df of $V_1^n,\ldots,V_n^n$. Then, using the equivalence $\bar{\psi}_\theta(t)\leq s \Leftrightarrow t\leq\psi_\theta(s)$ (componentwise) $\forall t\in E_k,\forall s\in\bar{\mathbb{R}}_k$, one easily obtains $C_n=\int[\bar{Z}_n(t)]^2\bar{F}(dt,\hat{\theta}_n)$ a.s. (F_θ) and $K_n=\|\bar{Z}_n\|_\infty$ a.s. (F_θ) with

(1.5) $\bar{Z}_n(t) = n^{1/2}(\bar{F}_n(t)-\bar{F}(t,\hat{\theta}_n))$, $t\in E_k$,

where $\|\cdot\|_\infty$ is the sup-norm for functions on E_k or $\mathbb{R}_k$. $\bar{Z}_n=(\bar{Z}_n(t),t\in E_k)$ is the "empirical process when parameters are estimated" appearing in the title of this paper. $\bar{Z}_n$ has paths belonging to the space $D_k=D[0,1]^k$ of functions on E_k having only jump discontinuities, see e.g. Neuhaus [13]. D_k furnished with a certain metric d of Skorohod-type becomes a Polish space. If $\mathcal{B}_d$ is the Borel-σ-algebra, $\bar{Z}_n$ is a rv with values in $(D_k, \mathcal{B}_d)$. In the subsequent sections we first prove weak convergence of the D_k-valued rv's $\bar{Z}_n, n \geq 1$, under H_o, and then extend this result to the whole class of contiguous alternatives. Such a procedure is characteristic to results concerning contiguous alternatives, and it is one of the main technical advantages of this concept that one can start investigations with the much simpler case where H_o holds true.

2. Convergence in distribution of the empirical process under the null hypothesis

Throughout in this section θ is fixed, and we write $F=F_\theta$, $\psi=\psi_\theta$, $\bar{\psi}=\bar{\psi}_\theta$ and $\bar{F}=\bar{F}(\cdot,\theta)$. Let us introduce the auxiliary process

(2.1) $\bar{X}_n(t)=n^{1/2}(\bar{F}_n(t)-\bar{F}(t))-n^{-1/2}\sum_{j=1}^{n} h(U_j^n,\theta)'q(t,\theta)$, $t\in E_k, n\geq 1$,

which is a rv with values in $(D_k,\mathcal{B}_d)$. In a first step we show

<u>Lemma 2.1.</u> Under the assumption of sec. 1 one has

$$\|\bar{Z}_n-\bar{X}_n\|_\infty \to o \text{ in } F\text{-probability.}$$

<u>Proof.</u> Define $\bar{R}_n=n^{1/2}(\bar{F}(\cdot,\theta)-\bar{F}(\cdot,\hat{\theta}_n))-n^{1/2}(\theta-\hat{\theta}_n)'q(\cdot,\theta)$, $n\geq 1$. Then apparently $\bar{Z}_n-\bar{X}_n=\bar{R}_n-R_n'q(\cdot,\theta)$, see (1.1). From $R_n=o_{F_\theta}(1)$ it follows that $\|R_n'q(\cdot,\theta)\|_\infty\to o$ in F_θ-probability. To handle $\bar{R}_n$, write with a suitable $\theta_n^*=\theta_n^*(t,\hat{\theta}_n)$ between θ and $\hat{\theta}_n$

(2.1) $\bar{R}_n(t)=n^{1/2}(\theta-\hat{\theta}_n)'(\partial F(\bar{\psi}(t),\theta_n^*)/\partial\theta-\partial F(\bar{\psi}(t),\theta)/\partial\theta)$

$=n^{1/2}(\theta-\hat{\theta}_n)'(q(\psi_{\theta_n^*}\circ\bar{\psi}(t),\theta_n^*)-q(t,\theta))$,

where the second equality follows from $\bar{\psi}_{\theta_n^*}\circ\psi_{\theta_n^*}\circ\bar{\psi}=\bar{\psi}$. On the other hand, if ι is the identity map on E_k one has

(2.2) $\|\psi_{\theta_n^*}\circ\bar{\psi}-\iota\|_\infty \leq \sup\{\|F(\cdot,\theta_1)-F(\cdot,\theta)\|_\infty:\theta_1\in\Theta, |\theta_1-\theta|\leq|\hat{\theta}_n-\theta|\}$,

and the RHS in (2.2) tends to zero in $F(\cdot,\theta)$-probability. Combined with (1.4) this yields $\|q_i(\psi_{\theta_n^*}\circ\bar{\psi},\theta_n^*)-q_i(\cdot,\theta)\|_\infty\to o$ in $F(\cdot,\theta)$-probability, $i=1,\dots,r$. An application of the last result to (2.1) entails $\|\bar{R}_n\|_\infty\to o$ in $F(\cdot,\theta)$-probability, since $n^{1/2}(\theta-\hat{\theta}_n)$ converges in distribution as (1.1) shows. The lemma is proved.┘

The lemma shows, that $\bar{X}_n, n\geq 1$, and $\bar{Z}_n, n\geq 1$, (under H_o) have the same limiting behaviour. Write $\bar{h}(\cdot)=h(\bar{\psi}(\cdot),\theta)$ and 1_t for the indicator function of the rectangle $(-\infty,t_1]\times\dots\times(-\infty,t_k]$, $t\in\mathbb{R}_k$. Then

(2.3) $\bar{g}(t,v) = 1_t(v)-\bar{F}(t)-\bar{h}(v)'q(t,\theta)$, $t,v\in E_k$,

defines a measurable function on $E_k\times E_k$ with the properties

(2.4) $\quad \bar{g}(\cdot,v)\in D_k, \forall v\in E_k;\ \int\bar{g}(t,\cdot)^2 d\bar{F}<\infty,\ \int\bar{g}(t,\cdot)d\bar{F}=0, \forall t\in E_k,$ and

we can write

(2.5) $\quad \bar{X}_n = n^{-1/2}\sum_{j=1}^{n}\bar{g}(\cdot,V_j^n)\ ,n\geq 1.$

Because of (2.4) $\bar{X}_n$ is centered with finite covariance kernel

(2.6) $$\bar{R}(t_1,t_2) = \int\bar{g}(t_1,\cdot)\bar{g}(t_2,\cdot)d\bar{F} =$$
$$= \bar{F}(t_1\wedge t_2)-\bar{F}(t_1)\cdot\bar{F}(t_2)-\langle 1_{t_1},\bar{h}\rangle' q(t_2,\theta)-\langle 1_{t_2},\bar{h}\rangle' q(t_1,\theta) +$$
$$+ q(t_1,\theta)'B(\theta)q(t_2,\theta)\quad ,\ t_1,t_2\in E_k,$$

with $t_1\wedge t_2$ the componentwise minimum of t_1,t_2 and $\langle 1_t,\bar{h}\rangle' = (\int 1_t\bar{h}_1 d\bar{F},\ldots,\int 1_t\bar{h}_r d\bar{F})$.

Theorem 2.2. Under the assumptions of section 1 there exists a centered Gaussian process $\bar{X}$ with covariance kernel (2.6) and paths belonging to the space $C(E_k)$ of continuous functions on E_k, such that under H_o

(2.7) $\quad \bar{X}_n \xrightarrow{\mathcal{D}} \bar{X}$ in D_k, and

(2.8) $\quad \bar{Z}_n \xrightarrow{\mathcal{D}} \bar{X}$ in D_k.

Proof. Because of Lemma 2.1 it is enough to show (2.7). The Cramêr-Wold device entails convergence of the finite dimensional distributions to those of a centered Gaussian process with covariance kernel (2.6). To prove tightness of the sequence $\bar{X}_n, n\geq 1$, write $\bar{X}_n = \sum_{i=o}^{r}\bar{X}_{in}$ with $\bar{X}_{on} = n^{1/2}(\bar{F}_n-\bar{F})$ and $\bar{X}_{in} = -(n^{-1/2}\sum_{j=1}^{n}\bar{h}_i(V_j^n,\theta))q_i(\cdot,\theta)$, $i=1,\ldots,r$. Then it is enough to prove for $i=o,1,\ldots,r$

(2.9) $$\lim_{\delta\to o}\limsup_{n\to\infty} \underset{\sim}{P}\{w(X_{in},\delta)\geq\varepsilon\} = o\ ,\forall\varepsilon>0,$$

with $w(f,\delta) = \sup\{|f(t_1)-f(t_2)|: t_1,t_2\in E_k,\ |t_1-t_2|<\delta\}, f\in D_k, \delta>o$, and moreover (2.9) entails that the limiting process $\bar{X}$ has paths in $C(E_k)$, see Neuhaus [13]. Since $\bar{F}$ is continuous, for $i=o$ (2.9) is wellknown to hold true, see e.g. Neuhaus [13]. For $1\leq i\leq r$ one has

(2.10) $$\underset{\sim}{P}\{w(\bar{X}_{in},\delta)\geq\varepsilon\} = \underset{\sim}{P}\{|n^{-1/2}\sum_{j=1}^{n}\bar{h}_i(V_j^n,\theta)|\,w(q_i(\cdot,\theta),\delta)\geq\varepsilon\}.$$

Since $n^{-1/2}\sum_{j=1}^{n}\bar{h}_i(V_j^n,\theta)$ converges in distribution and $q_i(\cdot,\theta)\in C(E_k)$,

(2.9) follows.┘

Remark 2.3. A k-dimensional version of a certain process $\tilde{Z}_n, n\geq 1$, used by Durbin [8] for k=1 may be defined by

$$\tilde{Z}_n(t) = n^{1/2}\{n^{-1}\sum_{j=1}^{n} 1_t(\psi_{\hat{\theta}_n}(U_j^n)) - F(\bar{\psi}_{\hat{\theta}_n}(t), \hat{\theta}_n)\}, \quad t\in E_k, n\geq 1.$$

Since all df's $F(\cdot,\theta)$ are continuous, it follows from (1.3)

(2.10) $\tilde{Z}_n = \bar{Z}_n(\psi\circ\bar{\psi}_{\hat{\theta}_n})$ a.s. (F_θ).

Let D_o be the subspace of D_1 consisting of the continuous df's in D_1 and D_o^k the k-fold product space of D_o. Then $\psi\circ\bar{\psi}_{\hat{\theta}_n}\in D_o^k$, and from $\hat{\theta}_n\to\theta$ in F_θ-probability one concludes

$$\|\psi\circ\bar{\psi}_{\hat{\theta}_n} - \iota\|_\infty \to o \quad \text{in } F_\theta\text{-probability.}$$

Therefore, in $D_k\times D_o^k$

(2.11) $(\bar{Z}_n, \psi\circ\bar{\psi}_{\hat{\theta}_n}) \overset{\mathcal{D}}{\to} (\bar{X}, \iota)$.

Generalizing the measurability considerations of Billingsley [4], p. 232, the map $T: D_k\times D_o^k \to D_k$ defined by $T(f,g)=f\circ g-f$ is (product-)measurable and continuous at the points of $C(E_k)\times D_o^k$. Consequently $T(\bar{Z}_n, \psi\circ\bar{\psi}_{\hat{\theta}_n}) \overset{\mathcal{D}}{\to} T(\bar{X},\iota)\equiv o$, i.e. $\|\tilde{Z}_n - \bar{Z}_n\|_\infty \to o$ in probability under H_o. Therefore $\tilde{Z}_n$ and $\bar{Z}_n$ are asymptotically (under H_o) identical.┘

Since $\|\cdot\|_\infty$ is continuous on $C(E_k)$ (2.8) immediately implies for the KS-statistic $K_n \overset{\mathcal{D}}{\to} \|\bar{X}\|_\infty$ under H_o. The parallel result for the CvM-statistic is not so immediate since there is not a single a.s. continuous function Λ on D_k with $C_n=\Lambda(\bar{Z}_n)$ but only a sequence $\Lambda_n=\int(\cdot)^2 d\bar{F}(\cdot,\hat{\theta}_n)$ which even depends on the observations. To overcome this difficulty let M_1 be the family of df's on E_k assigning mass o to the boundary of E_k. Then $M_1\subset D_k$, and with the Levy-Prohorov metric L, M_1 becomes a separable metric space, whose Borel-σ-algebra $\mathcal{B}(M_1)$ coincides with $D_k\cap\mathcal{B}_d$, see Neuhaus [15]. Let $\bar{W}_n=\bar{W}_n(V_1^n,\dots,V_n^n), n\geq 1$, be a sequence of rv's with values in M_1, and write $C(\bar{W}_n)=\int\bar{Z}_n^2(t)\bar{W}_n(dt)$. Then one has

Lemma 2.4. The assumptions of sec. 1 and

(2.12) $L(\bar{W}_n,\bar{W}) \to 0$ in F_θ-probability for some $\bar{W}\in M_1$

imply

(2.13) $C(\bar{W}_n) - \int \bar{X}_n^2 d\bar{W} \to 0$ in F_θ-probability,

and consequently

(2.14) $C(\bar{W}_n) \xrightarrow{\mathcal{D}} \int \bar{X}^2 d\bar{W}$.

<u>Proof.</u> From Lemma 3.1 we get first

(2.15) $C(\bar{W}_n) - \int \bar{X}_n^2 d\bar{W}_n \to 0$ in F_θ-probability,

and from (2.7)

$$(\bar{X}_n,\bar{W}_n) \xrightarrow{\mathcal{D}} (\bar{X},\bar{W}) \text{ on the product space } D_k \times M_1.$$

The function $T_1 : D_k \times M_1 \to \mathbb{R}$ defined by $T_1(f,G) = \int f^2 dG - \int f^2 d\bar{W}$ is $\mathcal{B}_d \otimes \mathcal{B}(M_1)$-$\mathcal{B}_1$ measurable and continuous at all points of $C(E_k) \times M_1$, see Neuhaus [15]. Therefore $T_1(\bar{X}_n,\bar{W}_n) \xrightarrow{\mathcal{D}} T_1(\bar{X},\bar{W}) \equiv 0$, i.e. $\int \bar{X}_n^2 d\bar{W}_n - \int \bar{X}_n^2 d\bar{W} \to 0$ in F_θ-probability; combined with (2.15) this yields (2.13).┘

With $\bar{W}_n = \bar{F}(\cdot,\hat{\theta}_n), n\geq 1$, and $\bar{W}=\bar{F}$ (2.12) is easily seen to be true, and one gets $C_n \xrightarrow{\mathcal{D}} \int \bar{X}^2 d\bar{F}$ under H_0. In Neuhaus [15] other possible choices of $\bar{W}_n, n\geq 1$ are discussed in detail.

3. Convergence in distribution of the empirical process under contiguous alternatives

Many of the considerations of this section are parallel to those in Neuhaus [14], sec. 2, where the L_2-approach was used. Therefore we can be short here.

From now on let $V_1^n,\ldots,V_n^n$ be i.i.d. with $\mathcal{L}(V_1^n)=P_n$ and corresponding df G_n, $G_n \notin H_0$, and assume that for some $\bar{F}=\bar{F}(\cdot,\theta)$ (θ fixed) the sequence $P_1^n=P_n\otimes\ldots\otimes P_n$, n-times, $n\geq 1$, is contiguous to $P_0^n=P_0\otimes\ldots\otimes P_0$, n-times, $n\geq 1$, where P_0 is the probability corresponding to $\bar{F}$. Contiguity is meant in the sense of Hâjek and Šidâk [9], p. 202. Let us start with special contiguous sequences $P_n, n\geq 1$, being dominated by P_0 with Radon-Nikodym (RN) derivatives

$$(3.1)\qquad dP_n/dP_0 = 1+n^{-1/2}a_n,\ a_n\in H = L_2(E_k,P_0), n\geq 1$$

with $||a_n-a|| \to 0$, $n\to\infty$, for some $a\in H$; $||\cdot||$ is the norm in H generated by the inner product $<\cdot,\cdot>$ in H.

The covariance kernel $\bar{R}(t_1,t_2)=<\bar{g}(t_2,\cdot),\bar{g}(t_1,\cdot)>$ from (2.6) is continuous on $E_k\times E_k$. Let $H(\bar{R})$ be the reproducing kernel Hilbert space of $\bar{R}$ consisting of continuous functions on E_k with characterizing properties: $\bar{R}(\cdot,t)\in H(\bar{R}), \forall t\in E_k$, and $<f,R(\cdot,t)>_{\bar{R}} = f(t), \forall t\in E_k$, where $<\cdot,\cdot>_{\bar{R}}$ is the inner product in $H(\bar{R})$ and $f\in H(\bar{R})$. Furthermore, we need a third Hilbert space, namely $H_0=L_2(D_k,\mathcal{B}_d,\bar{\mu}_0)$, with $\bar{\mu}_0=\mathcal{L}(\bar{X})$, see (2.7), and denote by $H(\bar{g})$ the closed linear subspace generated by $\bar{g}(t,\cdot), t\in E_k$, in H, and by $H_0(\pi)$ the closed linear subspace generated by the projections π_t, $t\in E_k$, in H_0, where $\pi_t(f)=f(t)$ for $f\in D_k$, $t\in E_k$. Then $H(\bar{R})$, $H(\bar{g})$ and $H_0(\pi)$ are isometrically isomorph as is indicated by

$$\begin{array}{ccccc} H(\bar{R}) & \overset{L^*}{\leftarrow} & H(\bar{g}) & \overset{L^0}{\rightarrow} & H_0(\pi) \\ \bar{R}(t,\cdot) & \leftarrow & \bar{g}(t,\cdot) & \rightarrow & \pi_t \end{array},$$

according to the "basic congruence theorem" of Parzen [18]. The following lemma takes the same place in the D_k-approach as Proposition 2.3 in Neuhaus [14] does in the L_2-approach.

Lemma 3.1. a) Let $\bar{x}_0$ be an element in D_k. Then the measures $\mathcal{L}(\bar{X})$ and $\mathcal{L}(\bar{X}+\bar{x}_0)$ on D_k are equivalent iff x_0 belongs to $H(\bar{R})$, i.e. iff $\bar{x}_0=L^*a$ for some $a\in H(\bar{g})$.

b) Write $\bar{\mu}_a=\mathcal{L}(\bar{X}+L^*a)$, $a\in H(\bar{g})$, and $\bar{\mu}_0=\mathcal{L}(\bar{X})$. Then one has

(3.2) $L_a = d\bar{\mu}_a/d\bar{\mu}_0 = \exp\,(Z_a-\|a\|^2/2)$,

where Z_a is a member of the equivalence class of $L^o a$.

c) If $b\in H(\bar{g})$ is a finite linear combination $b=\sum_{i=1}^{r}\alpha_i\bar{g}(t_i,\cdot)$ for some $t_i\in E_k,\alpha_i\in\mathbb{R}$, $i=1,\ldots,r,r\geq 1$, then Z_b may be written as $Z_b=\sum_{i=1}^{r}\alpha_i\pi_{t_i}$, i.e. Z_b is a.s. $(\bar{\mu}_0)$ continuous on D_k.

Proof. a) and b) are well-known, see e.g. Park [17]. c) follows from the linearity of L^o and from $\bar{\mu}_0(C(E_k))=1$.┘

For $a\in H(\bar{g})$ Z_a has normal distribution $\mathcal{N}(o,\|a\|^2)$ on $(D_k,\mathcal{B}_d,\bar{\mu}_0)$, and for $L_{n,b}=\exp\,\{n^{-1/2}\sum_{j=1}^{n} b(V_j^n)-\|a\|^2/2\}$, b as in Lemma 3.1. c),

(3.3) $L_b\circ\bar{X}_n = L_{n,b}$

holds true, since

$$Z_b\circ\bar{X}_n = \sum_{i=1}^{r}\alpha_i\bar{X}_n(t_i) = n^{-1/2}\sum_{j=1}^{n}\sum_{i=1}^{r}\alpha_i\bar{g}(t_i,V_j^n) = n^{-1/2}\sum_{j=1}^{n} b(V_j^n).$$

Our first result concerning convergence in distribution of $\bar{X}_n,n\geq 1$ under contiguous alternatives is

Theorem 3.2. Let K be a relatively compact subset of $H(\bar{g})$ and $A\in\mathcal{B}_d$ with $\bar{\mu}_0(\partial A)=o$, where ∂A is the boundary of A. Then under alternatives (3.1) for each $\varepsilon>o$ there are numbers $\delta>o$ and $n_0\leq 1$, such that

(3.4) $|P_1^n\{\bar{X}_n\in A\}-\bar{\mu}_a(A)|<\varepsilon$ holds true

for all pairs (a_n,a) with a_n as in (3.1), $a\in K,n\geq n_0$ and $\|a_n-a\|<\delta$. The proof of the above Theorems is the same as that of Theorem 2.4. in Neuhaus [14] after redefining the various quantities in an obvious manner and using Lemma 3.1 and (3.3).┘

The above theorem shows that for the special class of contiguous alternatives given by (3.1) the convergence in distribution is uniform in some sense. For the general class of contiguous alternatives which we now shall consider we can not derive such an uniform result, but convergence in distribution remains true.

Theorem 3.3. Assume that $P_1^n, n\geq 1$, is contiguous to $P_0^n, n\geq 1$. Then there exists a sequence $\bar{x}_n, n\geq 1$, in $C(E_k)$ such that

(3.5) $\bar{X}_n - \bar{x}_n \overset{\mathcal{D}}{\to} \bar{X}$ in D_k under the alternatives $P_n, n\geq 1$.

Proof. Let G_n be the df of P_n and write $W_n = n^{-1/2} \sum_{j=1}^{n} (1.(V_j^n) - G_n)$, $n\geq 1$. Then $\|G_n - \bar{F}\|_\infty \to o$ and $W_n \overset{\mathcal{D}}{\to} \bar{X}_o$ in D_k under $P_n, n\geq 1$, where $\bar{X}_o$ is a centered Gaussian process with continuous paths and covariance kernel $\bar{R}_o(t_1,t_2) = \bar{F}(t_1 \wedge t_2) - \bar{F}(t_1)\bar{F}(t_2)$, $t_1, t_2 \in E_k$. These results are well-known for $G_n \equiv \bar{F}$, see e.g. Neuhaus [13]. The general case then follows by a random change of time argument similar as in Remark (2.3). For sake of shortness we assume in the proof that $r=1$ (see (1.2)), i.e. $\bar{g}(t,v)$ has the form

(3.6) $\bar{g}(t,v) = 1_t(v) - \bar{F}(t) - \bar{h}(v)\cdot q(t)$, $t,v\in E_k$,

with $\bar{h}\in H$, $\int \bar{h}\, dP_o = o$, $q\in C(E_k)$.

Now, let us choose functions $\bar{h}_n \in H, n\geq 1$, with

(3.7) $\int \bar{h}_n dP_o = o, n\geq 1$; $\|\bar{h}_n - \bar{h}\| \to o$ and $n^{-1}\|\bar{h}_n\|_\infty^4 \to o$.

Then from Behnen and Neuhaus [3] one gets

(3.8) $H_n = n^{-1/2} \sum_{j=1}^{n} \bar{h}_n(V_j^n) - n^{1/2} \int \bar{h}_n dP_n \overset{\mathcal{D}}{\to} \mathcal{N}(o, \|\bar{h}\|^2)$ under

the alternatives $P_n, n\geq 1$. If $\bar{g}_n$ is defined as $\bar{g}$ in (3.6) with $\bar{h}$ replaced by $\bar{h}_n$ and $Y_n \equiv n^{-1/2} \sum_{j=1}^{n} \bar{g}_n(V_j^n)$ one notices that under H_o $E\|\bar{X}_n - Y_n\|_\infty^2 = \|\bar{h}_n - \bar{h}\|^2 \|q\|_\infty \to o$, and then contiguity entails

(3.9) $\|\bar{X}_n - Y_n\|_\infty \to o$ in P_n-probability.

It is therefore enough to show (3.5) with Y_n instead of $\bar{X}_n$. Write $\bar{x}_n = n^{1/2}(G_n - \bar{F} - \int \bar{h}_n dP_n \cdot q) \in C(E_k)$; then

(3.10) $Y_n - \bar{x}_n = W_n + q\cdot H_n$, $\forall n\geq 1$.

The relative compactness of $Y_n - \bar{x}_n, n \geq 1$, follows immediately from $W_n \xrightarrow{\mathcal{D}} \bar{X}_o$, (3.8), and (3.10). Furthermore, (3.7) and (3.10) entail

(3.11) $\quad \dot{Cov}(Y_n(t_1)-\bar{x}_n(t_1), Y_n(t_2)-\bar{x}_n(t_2)) \to \bar{R}(t_1,t_2)$ under $P_n, n \geq 1$.

Now an application of the Cramér-Wold-device and the Lindeberg-Theorem combined with (3.11) to the RHS of (3.10) yields the convergence of the finite-dimensional distributions of $Y_n - \bar{x}_n, n \geq 1$, to those of $\bar{X}$. The theorem follows.⌋

Corollary 3.4. If $P_n, n \geq 1$, fulfills (3.1), then Theorem 3.3 is true with $\bar{x}_n = L^* a$, i.e.

(3.12) $\quad \bar{X}_n \xrightarrow{\mathcal{D}} \bar{X} + L^* a$ in D_k under $P_n, n \geq 1$.

Proof. From the proof of Theorem 3.3 one has $\bar{x}_n = \int 1.a_n dP_o - \int a_n \bar{h}_n dP_o \cdot q, \forall n \geq 1$. Therefore $\|\bar{x}_n - L^* a\|_\infty \to 0$.⌋

The results of this paper can be used for studying the asymptotic power of Kolmogorov-Smirnov- and Cramér-von Mises-tests when parameters are present. Studies of this type for the CvM-test are made in Neuhaus [16], where it has turned out that there is great variation in the power for the various contiguous alternatives, and this justifies once more the use of this broad class of local nonparametric alternatives.

References

[1] Behnen, K. (1971). Asymptotic optimality and ARE of certain rank-order tests under contiguity. Ann. Math. Statist. 42 225-229.

[2] Behnen, K. (1972). A characterization of certain rank-order tests with bounds for the asymptotic relative efficiency. Ann.Math. Statist. 43 1839-1851.

[3] Behnen, K. and Neuhaus, G. (1975). A central limit theorem under contiguous alternatives. Ann. Statist. 3 1349-1353.

[4] Billingsley, P. (1968). Convergence of probability measures. J. Wiley, New York.

[5] Cramér, H. (1945). Mathematical methods of statistics. Almqvist and Wiksells, Uppsala. Princeton 1946.

[6] Darling, D.A. (1955). The Cramér-Smirnov test in the parametric case. Ann. Math. Statist. 26 1-20.

[7] Doob, J.L. (1949). Heuristic approach to the Kolmogorov-Smirnov theorems. Ann. Math. Statist. 20 393-403.

[8] Durbin, J. (1973). Weak convergence of the sample distribution function when parameters are estimated. Ann. Statist. 1 279-290.

[9] Hájek, J. and Šidák, Z. (1967). Theory of Rank Tests. Academic Press, New York.

[10] Kac, M. (1951). On some connections between probability theory and differential and integral equations. Proc. Second Berkeley Symp. Math. Statist. Probab., Univ. of Calif. Press 189-215.

[11] Kac, M. , Kiefer, J. and Wolfowitz, J. (1955) On tests of normality and other test of goodness of fit based on distance methods. Ann. Math. Statist. 26 189-211.

[12] Le Cam, L. (1960). Locally asymptotically normal families of distributions. Univ. of Calif.Publ. in Stat. 3 37-98.

[13] Neuhaus, G. (1971). On weak convergence of stochastic processes with multi-dimensional time parameter. Ann. Math. Statist. 42 1285-1295.

[14] Neuhaus, G. (1973). Asymptotic properties of the Cramér-von Mises statistic when parameters are estimated. Proc. Prague Symp. on Asymptotic Stat. Sept. 3-6, 1973 (J.Hájek,ed.) Universita Karlova Praha **2** 257-297.

[15] Neuhaus, G. (1973). Zur Verteilungskonvergenz einiger Varianten der Cramér-von Mises-Statistik. Math. Operationsforschung u. Statist. **4** 473-484.

[16] Neuhaus, G. (1976). Asymptotic power properties of the Cramér-von Mises test under contiguous alternatives. J. Multivariate Anal. **6** 95-110.

[17] Park, W.J. (1970). A multi-parameter Gaussian process. Ann. Math. Statist. **41** 1582-1595.

[18] Parzen, E. (1959). Statistical inference on time series by Hilbert space methods, I. Technical report No. 23, Department of Statistics, Stanford Univ..

[19] Sukhatme, S. (1972). Fredholm determinant of a positive kernel of a special type and its applications. Ann. Math. Statist. **43** 1914-1926.

Almost sure invariance principles for empirical distribution functions of weakly dependent random variables.

Walter Philipp

1. Introduction

By and large, sums of weakly dependent random variables such as mixing, lacunary trigonometric, Gaussian, etc. behave almost like sums of independent random variables. The literature provides many examples of this phenomenon. However, the situation may change drastically for empirical distribution functions. As an illustration consider a stationary sequence $\{\eta_n, n \geq 1\}$ of random variables uniformly distributed over $[0,1]$ with empirical distribution function $F_N(t)$ at stage N. Recall that $F_N(t) = F_N(t,\omega)$ is defined on $[0,1]$ as N^{-1} times the number of indices $n \leq N$ satisfying $\eta_n < t$. If the η_n's are independent, then according to a theorem of Donsker (1952) (see also Billingsley (1968) section 13) $N^{\frac{1}{2}}(F_N(t) - t)$ converges in distribution to the Brownian bridge over $[0,1]$. If, however, the η_n's are φ-mixing or lacunary then we have convergence in law to a certain Gaussian process, which, in general, is different from the Brownian bridge (see Billingsley (1968), section 22) and Billingsley (1967).

The functional law of the iterated logarithm is another example of this phenomenon. Let $D[0,1]$ be the space of functions on $[0,1]$ which are right continuous and have left-hand limits. Give D the topology defined by the supremum norm $\|\cdot\|_\infty$. For $N \geq 3$ put

$$(1.1) \qquad f_N(t) = N(F_N(t) - t)(2N \log\log N)^{-\frac{1}{2}}, \quad 0 \le t \le 1.$$

Let K be the set of all absolutely continuous functions h on $[0,1]$ with $h(0) = h(1) = 0$ and $\int_0^1 (dh/dt)^2 dt \le 1$. Then according to a theorem of Finkelstein (1971) the sequence $\{f_N(t), N \ge 3\}$ is with probability 1 relatively compact in $D[0,1]$ and has K as the set of its limit points. But if the random variables are m-dependent then according to a recent result of Oodaira (1975) the set of limit points is the unit ball in the reproducing kernel Hilbert space associated with the covariance function of the appropriate Gaussian limit process, which, in general, is different from the class K defined above. (For a definition of reproducing kernel Hilbert space see section 3.3 below.)

The purpose of this paper is to establish functional laws of the iterated logarithm for the empirical distribution functions of functions of random variables satisfying a strong mixing condition as well as for the empirical distribution functions of lacunary sequences $\{\langle n_k \omega\rangle, k \ge 1\}$. For random variables satisfying a strong mixing condition partial results have been obtained by Oodaira (1975). Furthermore, Oodaira in his paper points out that the most natural way to describe the limit points of sequences $\{f_N(t)\}$ for dependent random variables is in terms of the reproducing kernel Hilbert space.

In the lacunary case we obtain as a byproduct a result in probabilistic number theory on the discrepancy of lacunary sequences. Let $\{n_k, k \ge 1\}$ be a lacunary sequence of real numbers, i.e. a sequence satisfying

$$(1.2) \qquad n_{k+1}/n_k \ge q > 1$$

for all $k \ge 1$. Let $\{[0,1], \mathfrak{F}, \lambda\}$ be the unit interval with

Lebesgue measurability and Lebesgue measure λ. Then $\{\langle n_k\omega\rangle,\ k \geq 1\}$ can be considered as a sequence of random variables with asymptotically uniform distribution. Here $\langle \varepsilon \rangle$ denotes the fractional part of ε. Let $F_N(t)$ be the empirical distribution function at stage N. Then

$$(1.3) \qquad D_N = D_N(\omega) = \sup_{0\leq t\leq 1} |F_N(t) - t|$$

is called the discrepancy of the sequence $\{\langle n_k\omega\rangle,\ 1\leq k\leq N\}$, a concept important in probability as well as in number theory. Recently I proved (Philipp (1975)) that for lacunary sequences of integers

$$(1.4) \qquad \frac{1}{4} \leq \limsup_{N\to\infty} \frac{ND_N(\omega)}{\sqrt{N \log\log N}} \leq C(q)$$

with probability 1 where $C(q)$ is a constant depending on q only. The right-hand inequality in (1.4) was conjectured by Erdős and Gaal in 1954 (see Erdős (1964), p. 56). In this paper it is shown that (1.4) continues to hold for lacunary sequences $\{n_k\}$ which are not necessarily integer.

Except for the value of the constant, the left-hand inequality in (1.4) was well known since the publication of a result of Erdős and Gál (1955). As a matter of fact, this left inequality was the basis for their conjecture. For a proof of the left inequality and a short history of the conjecture see Philipp (1975).

In the recent past rather efficient methods have been developed to treat sums of weakly dependent random variables. All of these methods rely on some kind of approximation scheme for dependent random variables by a martingale difference sequence. These methods are quite powerful since all the heavy machinery for martingales is then at one's disposal. (For an extensive account see e.g. the recent memoir by Philipp and Stout (1975).)

In this paper another kind of martingale approximation is used which is simpler and more easily applicable than all the previous ones.

2. Statement of results.

2.1 Functions of strongly mixing random variables

Let $\{\xi_n, n \geq 1\}$ be a strictly stationary sequence of random variables satisfying a strong mixing condition

$$|P(AB) - P(A)P(B)| \leq \rho(n) \tag{2.1.1}$$

for all $A \in \mathfrak{F}_1^t$ and $B \in \mathfrak{F}_{t+n}^{\infty}$. Here $\mathfrak{F}_a^b$ denotes the σ-field generated by ξ_n $(a \leq n \leq b)$. Let f be a measurable mapping from the space of infinite sequences $(\alpha_1, \alpha_2, \dots)$ of real numbers into the real line. Define

$$\eta_n = f(\xi_n, \xi_{n+1}, \dots), \quad n \geq 1 \tag{2.1.2}$$

and

$$\eta_{mn} = E(\eta_n | \mathfrak{F}_n^{n+m}), \quad m,n \geq 1. \tag{2.1.3}$$

As is usual we assume that η_n can be closely approximated by η_{mn} in the form

$$E|\eta_n - \eta_{mn}| \leq \psi(m) \downarrow 0 \tag{2.1.4}$$

for all $m,n \geq 1$.

Denote by $F_N(t)$ the empirical distribution function of the sequence $\{\eta_n, n \geq 1\}$ at stage N. We assume that η_n is uniformly distributed over $[0,1]$. Write

$$f_N(t) = N(F_N(t) - t)(2N \operatorname{loglog} N)^{-\frac{1}{2}}, \quad 0 \leq t \leq 1. \tag{2.1.5}$$

Theorem 2.1. Let $\{\xi_n, n \geq 1\}$ be a strictly stationary sequence of random variables satisfying a strong mixing condition (2.1.1) with

$$\rho(n) \ll n^{-8}. \tag{2.1.6}$$

(Throughout the Vinogradov symbol $\ll$ instead of O is used whenever convenient.)

Suppose that the random variables η_n defined by (2.1.2) are uniformly distributed over $[0,1]$ and that they satisfy (2.1.4) with

$$\psi(m) \ll m^{-12}. \tag{2.1.7}$$

Then for each $\varepsilon > 0$ there is with probability 1 a random index $N_0 = N_0(\varepsilon)$ such that

$$|f_N(t) - f_N(s)| \leq C|t-s|^{1/120} + \varepsilon \tag{2.1.8}$$

for all $0 \leq s \leq t \leq 1$ and all $N \geq N_0$. The constant C only depends on the constants implied by $\ll$ in (2.1.6) and (2.1.7). In particular (2.1.8) implies that the sequence $\{f_N(t), N \geq 3\}$ is with probability 1 relative compact in $D[0,1]$.

In order to identify the limits of the sequence $\{f_N(t)\}$ we need some more notation and an additional hypothesis. Write

$$g_n(t) = 1\{0 \leq \eta_n < t\} - t = x_n(0,t). \tag{2.1.9}$$

Under the hypothesis of theorem 2.1 the two series defining the covariance function

$$\Gamma(s,t) = E(g_1(s)g_1(t)) + \sum_{n=2}^{\infty} E(g_1(s)g_n(t)) + \sum_{n=2}^{\infty} E(g_n(s)g_1(t)) \tag{2.1.10}$$

$(0 \leq s,t \leq 1)$ converge absolutely (see Billingsley (1968) section 22).

Let $\{T_m, m \geq 1\}$ be an increasing sequence of finite subsets $\{t_1, \ldots, t_m\} \subset [0,1]$ such that $\bigcup_{m>1} T_m$ is dense in $[0,1]$. Let B_m be the set of all functions f on $[0,1]$ defined by

$$f(x) = \sum_{j \leq m} \alpha_j \Gamma(x,t_j), \qquad \alpha_j \in \mathbb{R}$$

satisfying

$$\sum_{j,k\leq m} \alpha_j \alpha_k \, \Gamma(t_j, t_k) \leq 1.$$

Theorem 2.2. Suppose that in addition to the hypothese of Theorem 2.1 the covariance function $\Gamma(s,t)$ is positive definite. Then the sequence $\{f_N(t),\ N \geq 3\}$ has with probability 1 the unit ball in the reproducing kernel Hilbert space $H(\Gamma)$ as its set of limit points. Equivalently, the set of limit points equals $\overline{\bigcup_{m\geq 1} B_m}$ where the closure is in the topology defined by the supremum norm over [0,1].

Remarks. (2.1.8) implies

(2.1.11) $$\limsup_{N\to\infty} \sup_{0\leq t\leq 1} |f_N(t)| \leq C \qquad \text{a.s.}$$

Except for the value of the constant (2.1.11) can be regarded as a generalization of the Chung-Smirnov law of the iterated logarithm for empirical distribution functions of independent uniformly distributed random variables (see Chung (1949)). For independent random variables Cassels (1951) proved that (2.1.8) holds with the right-hand side replaced by $((t-s)(1-t+s))^{\frac{1}{2}} + \varepsilon$. Hence except for the value of C relation (2.1.8) applied to independent random variables stands somewhere between Cassels' theorem and the Chung-Smirnov theorem.

But actually much more is true. We first observe that Theorem 2.2 contains Finkelstein's result as a special case since, as is well-known, the limit set appearing in Finkelstein's (1971) Theorem 1 is precisely the unit ball in the reproducing kernel Hilbert space of the Brownian bridge.

Second it might be interestirg to note that Finkelstein's theorem (and hence Theorem 2.2) implies Cassel's theorem .

To prove this we need the following lemma, due to Riesz (1955), p. 75.

Lemma 2.1. Let f be a real valued function on the unit interval. The following two conditions are equivalent:

1.) f is absolutely continuous with respect to Lebesgue measure and

$$\int_0^1 (f'(x))^2 \, dx \leq 1$$

2.) For every finite partition $0 \leq x_0 < x_1 < \ldots < x_s \leq 1$ of $[0,1]$,

$$\sum_{i=1}^{s} \frac{(f(x_i) - f(x_{i-1}))^2}{x_i - x_{i-1}} \leq 1.$$

We now shall prove that Finkelstein's theorem implies Cassels' theorem which in turn obviously implies the Chung-Smirnov law of the iterated logarithm. Indeed, by lemma 2.1 we observe that for each function $f \in K$ we have for $0 \leq s < t \leq 1$

$$\frac{f^2(s)}{s} + \frac{(f(t) - f(s))^2}{t-s} + \frac{f^2(t)}{1-t} \leq 1.$$

Since by elementary calculations

$$\frac{f^2(s)}{s} + \frac{f^2(t)}{1-t} \geq \frac{(f(t) - f(s))^2}{1-t+s}$$

we conclude that

$$|f(t) - f(s)| \leq \{(t-s)(1-t+s)\}^{\frac{1}{2}} .$$

Using the relative compactness of $\{f_N(t), N \geq 3\}$ one can now easily deduce Cassel's theorem.

We shall show now that (2.1.8) implies the relative compactness of $\{f_N(t),\ N \geq 3\}$ over $[0,1]$. In order to apply the Arzelà-Ascoli theorem we approximate $f_N(t)$ by a continuous $h_N(t)$ as follows. Fix $\omega \in \Omega_1$, where Ω_1 is the set on which (2.1.8) holds. Denote by $\alpha_1,\dots,\alpha_M$ the discontinuities of $f_N(t)$, $0 \leq t \leq 1$ and put $\alpha_0 = 0$ and $\alpha_{M+1} = 1$. We define $h_N(t)$ to be a piecewise linear function on $[0,1]$ with

$$h_N(\alpha_m) = f_N(\alpha_m) \qquad 0 \leq m \leq M+1$$

By comparing the graphs of h_N and f_N we observe that on each interval $(\alpha_m, \alpha_{m+1}]$

$$0 \leq f_N(t) - h_N(t) \leq f_N(\alpha_m+) - f_N(\alpha_m) \leq 2\varepsilon \tag{2.1.12}$$

for $N \geq N_0$ using (2.1.8). Let $0 \leq s < t \leq 1$ with $C|t-s|^{1/120} < \varepsilon$. Then by (2.1.12) and (2.1.8)

$$|h_N(s) - h_N(t)| < 5\varepsilon$$

for $N \geq N_0$. Hence $\{h_N(t),\ N \geq 3\}$ is equicontinuous over $[0,1]$. Moreover, it is uniformly bounded since $\{f_N(t),\ N \geq 3\}$ is. Thus by the Arzelà-Ascoli theorem $\{h_N(t),\ N \geq 3\}$ is relatively compact over $[0,1]$ and so is $\{f_N(t),\ N \geq 3\}$ by (2.1.12).

2.2 Lacunary sequences

Let $\{n_k, k \geq 1\}$ be a sequence of real numbers satisfying

$$n_{k+1}/n_k \geq q > 1 \quad (k = 1,2, \dots)$$

for some $q > 1$. For fixed s and t with $0 \leq s < t \leq 1$ write $L = [s,t)$, $\ell = t - s$ and

$$(2.2.1) \qquad x_k = x_k(s,t) = 1\{s \leq n_k\omega < t\} - (t-s) = 1_L(n_k\omega) - \ell$$

where $1\{\dots\} = 1_L\{.\}$ is extended with period 1. In other words we are investigating the sequence $\{\langle n_k\omega\rangle, k \geq 1\}$ of random variables as described in section 1. Denote by $F_N(t)$ the empirical distribution function of $\{\langle n_k\omega\rangle, k \geq 1\}$ at stage N. Define

$$(2.2.2) \qquad f_N(t) = N(F_N(t) - t)(2N \log\log N)^{-\frac{1}{2}} \quad (0 \leq t \leq 1)$$

Theorem 2.3. Let $\{n_k, k \geq 1\}$ be a lacunary sequence of real numbers. Then for each $\varepsilon > 0$ there exists with probability 1 a $N_0(\varepsilon)$ such that

$$(2.2.3) \qquad |f_N(t) - f_N(s)| \leq C|t - s|^{\frac{1}{4}} + \varepsilon$$

for all $N \geq N_0$ and all $0 \leq s < t \leq 1$. The constant C only depends on q. In particular, (2.2.3) implies that the sequence $\{f_N(t), N \geq 3\}$ is relatively compact in $D[0,1]$.

The statement about the relative compactness can be shown as in section 2.1.

As pointed out in section 1 Theorem 2.3 also implies a law of the iterated logarithm of the form (1.4). Indeed, we have with probability 1

$$\frac{N|F_N(t) - t|}{\sqrt{N \log\log N}} \ll 1$$

for all $N \geq N_0$ and for all $0 \leq t \leq 1$. Hence taking first the supremum over all t with $0 \leq t \leq 1$ and then the limes superior as $N \to \infty$ we obtain the right-hand side of (1.4).

We can identify the limit points of $\{f_N(t), N \geq 3\}$ only if we make some further assumptions. We assume that for all step functions with period 1 and $\int_0^1 f(x)dx = 0$ and for all $k \geq 1$, $0 \leq i < 2^k$ and $M, N \geq 1$ we have

(2.2.4) $$2^k \int_{i2^{-k}}^{(i+1)2^{-k}} \Big(\sum_{j=M+1}^{M+N} f(n_j\omega)\Big)^2 \, d\omega = \sigma^2 N + O\Big(\frac{N2^n}{n_M}\Big)$$

where σ and the constant implied by O depend on q and on f only. In particular, (2.2.4) implies that

(2.2.5) $$\lim_{N\to\infty} \frac{1}{N} E\Big(\sum_{j,k\leq N} x_j(0,s)x_k(0,t)\Big) = \Gamma(s,t)$$

exists.

Of course (2.2.4) says that the conditional variances of the partial sums given the σ-field generated by the dyadic intervals of order k equal asymptotically the variances of these sums which in turn equal asymptotically a constant multiple of the length of these partial sums. Under these hypotheses Berkes (1976) proved an almost sure invariance principle for the sums $\sum_{n\leq N} f(n_k\omega)$.

<u>Theorem 2.4</u>. Let $\{n_k, k \geq 1\}$ be a lacunary sequence of real numbers such that (2.2.4) is satisfied for all $0 \leq s < t \leq 1$. Suppose that $\Gamma(s,t)$ defined in (2.2.5) is positive definite. Then the sequence $\{f_N(t), N \geq 3\}$ is relatively compact in $D[0,1]$ and has the unit ball in RKHS $H(\Gamma)$ as the set of its limits points.

Equivalently, the set of limits points equals $\overline{\bigcup_{m\geq 1} B_m}$ where the closure is in the topology defined by the supremum norm over [0,1]. Here B_m is defined in the same way as before.

3. Description of the method and basic theorems

3.1 Chover's approach and Oodaira's proposition

Chover (1967) gave a proof of a weaker version of Strassen's (1964) functional law of the iterated logarithm for sums of independent identically distributed random variables using only classical results such as maximal inequalities and the central limit theorem with remainder. His approach consists of two steps. He first proves that the sequence of bookkeeping functions is with probability 1 uniformly equicontinuous and bounded and thus by the Arzelà-Ascoli theorem is relatively compact. He then identifies the class of limit points by showing that certain polygonal functions defined in terms of these bookkeeping functions converge to the corresponding polygonal functions defined in terms of Strassen's class K.

Chover's approach has proved to be useful in a variety of situations. A modified version of it, which at the same time is more general, has been formulated by Oodaira (1975). In order to describe it let us introduce some notation. Let $T = T_m = \{t_1, \ldots, t_m\}$ be a finite subset of $[0,1]$. Denote by $\phi^T = (\phi(t_1), \ldots, \phi(t_m))$ the restriction of a function ϕ to T and for a class A of functions ϕ on $[0,1]$ denote by $A^T = \{\phi^T : \phi \in A\}$. Let $\{T_m\}$ be an increasing sequence of subsets T_m such that $\bigcup_{m=1}^{\infty} T_m$ is dense in $[0,1]$. The following proposition is due to Oodaira (1975).

<u>Proposition 3.1</u>. Let $\{g_N(t) = g_N(t,\omega),\ N \geq 1\}$ to be a sequence of random functions in $C[0,1]$. Suppose that

(3.1.1) $\{g_N(t)\}$ is with probability 1 relatively compact

and that

(3.1.2) for each $T \in \{T_m\}$, the set of limit points of random vectors $\{g_N^T\}$ is K^T with probability 1 where K is a compact set in $C[0,1]$.

Then the set of limit points of $\{g_N(t)\}$ is K a.s.

Remark. Because of the Arzelà-Ascoli theorem (3.1.1) is equivalent with

(3.1.1') $\{g_N(t)\}$ is with probability 1 uniformly equicontinuous and bounded.

Hence in view of Oodaira's proposition the proof of the functional law of the iterated logarithm may be carried out in two steps, consisting of the proof of (3.1.2) and (3.1.1) or (3.1.1'). For our purposes (3.1.1') is more convenient to check.

It might be interesting to observe that Oodaira's proposition can be considered as an anologue to a well known theorem on weak convergence of probability measures used to prove distribution type invariance principles.

3.2 Relative compactness

Let $\{\eta_n, n \geq 1\}$ be a sequence of random variables with η_n uniformly distributed over $[0,1]$, and satisfying the hypotheses of Theorem 2.1. For fixed s and t with $0 \leq s < t \leq 1$ write

(3.2.1) $$L = [s,t), \quad \ell = t - s$$

and

(3.2.2) $$x_n = x_n(s,t) = 1_L(\eta_n) - \ell$$

We shall sketch the proof of probability estimates of the large deviations of the sums $\sum_{n=H+1}^{H+N} x_n$ for all $H \geq 0$ and $N \geq 1$. These estimates will then be used to prove the relative compactness of the random functions $\{f_N(t)\}$ defined by (1.1). For the sake of simplicity we assume that $f(\alpha_1, \alpha_2, \ldots) = \alpha_1$ so that (2.1.2) reduces to

$$(2.1.2') \qquad \eta_n = \xi_n.$$

As indicated in section 1 we shall approximate the above sums by martingales. For simplicity we consider the case $H = 0$ only. As a matter of fact it will turn out that there is no loss of generality in making this assumption. The dependence of the random variables η_n is such that the x_n's are also weakly dependent. In essence this means that

$$(3.2.3) \qquad E|E\{x_{n+k}|x_1,\ldots,x_n\}| \longrightarrow 0$$

as $k \longrightarrow \infty$ for each $n = 1,2,\ldots$. We construct inductively two sequences of progressively larger blocks H_j and I_j of consecutive integers. To fix the ideas let H_j consist of $[j^{\frac{1}{2}}]$ consecutive integers and let I_j also consist $[j^{\frac{1}{2}}]$ consecutive integers leaving no gaps between the blocks. The order is $H_1, I_1, H_2, I_2, \ldots$. Hence $H_1 = \{1\}$, $I_1 = \{2\}$, $H_2 = \{3\}$, $I_2 = \{4\}, \ldots, H_4 = \{7,8\}$, $I_4 = \{9,10\}, \ldots$.

We now define new random variables y_j and z_j by

$$y_j = \sum_{\nu \in H_j} x_\nu, \qquad z_j = \sum_{\nu \in I_j} x_\nu .$$

Of course, the size of the blocks can be adapted to the situation considered. For example, in other instances such as in the almost sure invariance principles treated by Philipp and Stout (1975) it is important that the number of elements in the blocks I_j is much smaller than the one in the corresponding block H_j. For example, card $I_j = [j^{\frac{1}{4}}]$ might be a reasonable choice since then the influence of the random variables z_j becomes progressively negligible and thus the z_j's can be discarded without doing any harm.

We now center the y_j at conditional expectations. (Of course, this technique dates back to P. Lévy.) So let $\mathfrak{L}_j$ be the σ-field generated by $y_1, \ldots, y_j$. Define

$$(3.2.4) \qquad Y_j = y_j - E(y_j | \mathfrak{L}_{j-1}).$$

Then $\{Y_j, \mathfrak{L}_j, j \geq 1\}$ is a martingale difference sequence. Moreover, since y_j is separated from $\mathfrak{L}_{j-1}$ by about $j^{\frac{1}{2}}$ indices we obtain

$$(3.2.5) \qquad E|E(y_j | \mathfrak{L}_{j-1})| \longrightarrow 0$$

as $j \longrightarrow \infty$ since the convergence in (2.1.6) to zero is sufficiently fast. Hence (3.2.4) shows that $\{y_j\}$ practically equals a martingale difference sequence $\{Y_j, \mathfrak{L}_j\}$. The sequence $\{z_j\}$ can be treated in a similar fashion.

We then apply the following exponential bound for martingales due to W. Stout (1974).

<u>Lemma 3.2.1.</u> Let $\{U_n, \mathfrak{F}_n\}_{n=1}^{\infty}$ be a supermartingale with $EU_1 = 0$. Put

$$U_0 = 0 \quad \text{and} \quad Y_j = U_j - U_{j-1} \qquad j \geq 1$$

Suppose that

$$Y_j \leq c \qquad \text{a.s.}$$

for some constant $c > 0$ and for all $j \geq 1$. For $\lambda > 0$ define

$$T_n = \exp\{\lambda U_n - \tfrac{1}{2} \lambda^2 (1 + \tfrac{1}{2} \lambda c) \sum_{j \leq n} E(Y_j^2 | \mathfrak{F}_{j-1})\}, \ n \geq 1$$

and $T_0 = 1$ a.s. Then for each λ with $\lambda c \leq 1$ the sequence $\{T_n, \mathfrak{F}_n\}_{n=1}^{\infty}$ is a nonnegative supermartingale satisfying

$$P\{\sup_{n\geq 0} T_n > \alpha\} \leq 1/\alpha$$

for each $\alpha > 0$.

This lemma is then used to obtain the following exponential bound.

Proposition 3.2.1. Let $H \geq 0$, $N \geq 1$ be integers and let $R \geq 1$. Suppose that $\ell \geq N^{-\frac{1}{2}}$ and that the hypotheses of Theorem 3.1 are satisfied. Then as $N \longrightarrow \infty$

$$P\{|\sum_{n=H+1}^{H+N} x_n| \geq AR\ell^{1/120}(N \log\log N)^{\frac{1}{2}}\}$$

$$\ll \exp(-6R\ell^{-1/120}\log\log N) + R^{-2}N^{-7/6}$$

where $A > 1$ and the constant implied by $<$ only depend on the constants implied by (2.1.6) and (2.1.7).

In the course of the proof we also need the following estimate which we state as a lemma.

Lemma 3.2.2. There is a constant $B \geq 1$ such that

$$P\{\sum_{j\leq M} E(Y_j^2|\mathfrak{L}_{j-1}) \geq 2RB\ell^{1/40}N\} \ll R^{-2}\,N^{-7/6}.$$

As was proved in section 2.1 relation (2.1.8) implies relative compactness. Now (2.1.8) follows at once from Proposition 3.2.1 and the following proposition which we state in full generality.

Proposition 3.2.2. Let $A \geq 1$, $\alpha > 0$ and $0 < \beta \leq 1$ be constants. Let $x_n = x_n(s,t)$ be defined by (3.2.2) for some sequence of random variables η_n. Suppose that

$$P\{|\sum_{n=H+1}^{H+N} x_n(s,t)| \geq AR\ell^{\alpha}(N \log\log N)^{\frac{1}{2}}\} \ll \exp(-3R\ell^{-\alpha}\log\log N) + R^{-2}N^{-1-\beta}$$

uniformly for all $H \geq 0$, $N \geq 1$, $R \geq 1$ and (s,t) with $0 \leq s < t \leq 1$ and $t-s \geq N^{-\frac{1}{2}}$. Then for each $\varepsilon > 0$ there exists with probablity 1 a $N_0(\varepsilon)$ such that

$$|\sum_{n<N} x_n(s,t)| \leq C(A,\alpha,\beta)((t-s)^{\alpha}+\varepsilon)(N \log\log N)^{\frac{1}{2}}$$

for all $N \geq N_0$ and all $0 \leq s < t \leq 1$. Here the constant $C(A,\alpha,\beta)$ depends on A, α and β only.

For the proof of Proposition 3.2.2 we use a triple dyadic expansion of $\sum_{n\leq N} x_n(s,t)$. More specifically we write N and both s and t in binary expansion and then sum over all the parameters. The details are somewhat intricate and are therefore omitted.

The characterization of the limit functions is given in section 3.4. We postpone its informal discussion until after the next section.

3.3 Reproducing kernel Hilbert spaces

We consider only the basic definitions in the simplest setup. Let $\Gamma(s,t)$ be a positive definite function on $E \times E$ where $E \subset \mathbb{R}$. Let K_m be the class of functions on E which can be written in the form

$$f(x) = \sum_{i\leq m} \alpha_i \Gamma(x,y_i) \tag{3.3.1}$$

where $y_i \in E$ and $\alpha_i \in \mathbb{R}$. (Not all α_i need be $\neq 0$). If

$$g(x) = \sum_{i\leq m} \beta_i \Gamma(x,y_i) \tag{3.3.2}$$

then the inner product (f,g) of f and g is defined by

(3.3.3) $$(f,g) = \sum_{j,k\leq m} \alpha_j\beta_k\ \Gamma(y_j,y_k)\ .$$

$\Gamma(s,t)$ has the reproducing kernel property on K_m since

(3.3.4) $$(f,\Gamma(\cdot,y_k)) = \sum_{j\leq m} \alpha_j\ \Gamma(y_j,y_k) = f(y_k).$$

This follows from (3.3.1) and (3.3.3) if we set in (3.3.2) $\beta_k = 1$ and $\beta_i = 0$ for $i \neq k$. The inner product defines a metric on $\bigcup_{m\geq 1} K_m$. (Note that we may have to set some of the α_i or $\beta_i = 0$ in order to obtain the same sets $\{y_\nu\}$ for both f and g.) But $\bigcup_{m\geq 1} K_m$ is, in general, not complete. The reproducing kernel Hilbert space $H(\Gamma)$ over E associated with $\Gamma(s,t)$ is then defined as the completion of $\bigcup_{m\geq 1} K_m$. Denote its norm by $\|\cdot\|_H$.

Let $T = \{t_1,\dots,t_m\}$ and let Γ^T denote the restriction of Γ to $T\times T$. Denote by $H(\Gamma^T)$ the reproducing kernel Hilbert space with reproducing kernel Γ^T.

Lemma 3.3.1 (Oodaira (1974)). For each T, the restriction of the unit ball of $H(\Gamma)$ to T is the unit ball of $H(\Gamma^T)$.

For more details on reproducing kernel Hilbert space see Aronszajn (1950) or Meschkowski (1962).

3.4 Identification of the limits

As pointed out in section 3.1 the second step in the proof of the functional law of the iterated logarithm consists of verifying condition (3.1.2) with $K = H(\Gamma)$. To this end we define random vectors $y_k \in \mathbb{R}^m$ with components $x_k(0,t_j)$ $(1 \leq j \leq m)$. Under the assumptions we made the $m\times m$ matrix $\Gamma_m = ((\Gamma(t_i,t_j)))_{i,j=1}^m$ defined by

$$\Gamma(t_i,t_j) = \lim_{N\to\infty} N^{-1} \sum_{\ell,k\leq N} E(x_k(0,t_i)x_\ell(0,t_j))$$

is positive definite. It then turns out that the sequence

$$\left\{ \frac{\sum_{k\leq N} y_k}{\sqrt{2N \log\log N}}, \quad N \geq 3 \right\}$$

of random vectors $\in \mathbb{R}^m$ is bounded almost surely and has the ellipsoid $E_m = \{x \in \mathbb{R}^m : x' \Gamma_m^{-1} x \leq 1\}$ as its set of limit points. This is proved by basic linear algebra, by means of a lemma reminiscent of the Cramer Wold device coupled with a law of the iterated logarithm for partial sums of weakly dependent random variables. By a simple linear transformation it is then shown that E_m equals the unit ball in the reproducing kernel Hilbert space $H(\Gamma_m)$. An application of lemma 3.3.1 will then show that (3.1.2) holds.

The following result which includes Theorem 2.2 was recently obtained jointly with I. Berkes.

Theorem 3.1. Let $\{\xi_k, k \geq 1\}$ be a strictly stationary sequence of random variables uniformly distributed over $[0,1]$ satisfying a strong mixing condition with

$$\rho(n) \ll n^{-8} .$$

Let $F_N(s) = F_N(s,\omega)$ $(0 \leq s \leq 1)$ be the empirical distribution function of $\xi_1(\omega), \ldots, \xi_N(\omega)$ and let

$$R(s,t) = t(F_{[t]}(s) - s) \qquad 0 \leq s \leq 1, \quad t \geq 0,$$

by the empirical process of the sequence. Let

$$g_s(\alpha) = 1_{[0,s]}(\alpha) - s$$

and suppose that

$$\Gamma(s,s') = E\{g_s(\xi_1)g_{s'}(\xi_1)\} + \sum_{k=1}^{\infty} E\{g_s(\xi_1)g_{s'}(\xi_k) + g_{s'}(\xi_1)g_s(\xi_k)\}$$

the covariance function of the corresponding Gaussian process is positive definite. Then, without changing the distribution of the process $\{R(s,t),\ 0 \le s \le 1,\ t \ge 0\}$, we can redefine $R(s,t)$ on a richer probability space together with a Kiefer-process $K(s,t)$ such that

$$\sup_{0 \le s \le 1} |R(s,t) - K(s,t)| \ll t^{\frac{1}{2}}/\log^{\alpha} t \qquad \text{a.s.}$$

for some $\alpha > 0$.

<u>Definition</u>. A separable Gaussian process $K(s,t)$ defined on $[0,1] \times [0,\infty)$ is called a Kiefer process if

$$E(K(s,t)) = 0$$

and

$$E\{K(s_1 . t_1) K(s_2,t_2)\} = \min(t_1,t_2) \cdot \Gamma(s_1,s_2) \ .$$

References

Aronszajn, N. (1950), The theory of reproducing kernels, Trans. Amer. Math. Soc. 68, 337-404.

Berkes, István (1975), On the asymptotic behavior of $\sum f(n_k,x)$ parts I + II, Z. Wahrscheinl, verw. Geb. 34, 319-365.

Billingsley, Patrick (1967), unpublished manuscript.

Billingsley, Patrick (1968), Convergence of probability measures, Wiley, New York.

Cassels, J. W. S. (1951), An extension of the law of the iterated logarithm, Proc. Cambridge Philos. Soc. 47, 55-64.

Chover, J. (1967), On Strassen's version of the log log law, Z. Wahrscheinl. verw. Geb. 8, 83-90.

Chung, K.-L. (1949), An estimate concerning the Kolmogoroff limit distribution, Trans. Amer. Math. Soc. 67, 36-50.

Donsker, M. (1952), Justification and extension of Doob's heuristic approach to the Kolmogorov-Smirnov theorems, Ann. Math. Stat. 23, 277-281.

Erdős, P. (1964), Problems and results on diophantine approximations, Compositio Math 16, 52-65.

Erdős, P. and Gál, I. S., On the law of the iterated logarithm, Proc. Koninkl. Nederl. Akad Wetensch. Ser. A.58, 65-84.

Finkelstein, Helen (1971), The law of the iterated logarithm for empirical distributions, Ann. Math. Stat. 42, 607-615.

Meschkowski, Herbert (1962), Hilbertsche Räume mit Kernfunktionen, Springer, Berlin.

Oodaira, Hiroshi (1975), Some functional laws of the iterated logarithm for dependent random variables, Colloqu. Math. Soc. Janos Bolyai 11, 253-272.

Philipp, Walter (1975), Limit theorems for lacunary series and uniform distribution mod 1, Acta Arithmetica 26, 241-251.

Philipp, Walter and Stout, William (1975), Almost sure invariance principles for partial sums of weakly dependent random variables, Memoirs Amer. Math. Soc. 161.

Riesz, F. and Sz. Nagy, B. (1955), Functional analysis, Frederick Unger, New York.

Smirnoff, N. (1939), Sur les ecarts de le courbe de distribution coupirique. Mat. Sbornik 6, 3-26.

Stout, W. F. (1974), Almost sure convergence. Academic Press, New York.

Strassen, V. (1965), An invariance principle for the law of the iterated logarithm. Z. Wahrscheinl. 3, 211-226.

Walter Philipp
Department of Mathematics
University of Illinois
Urbana, Illinois 61801

THREE THEOREMS OF MULTIVARIATE EMPIRICAL PROCESS

P. Révész

SUMMARY

Let $X_1, X_2, \ldots$ be a sequence of independent r.v.'s uniformly distributed over the unit cube I^d of the d-dimensional Euclidean Space. Further let F_n be the empirical distribution function based on the sample $X_1, X_2, \ldots, X_n$ and let

$$\alpha_n(x) = n^{1/2}(F_n(x) - x_1 x_2 \ldots x_d) \quad (x = (x_1, x_2, \ldots, x_d) \in I^d)$$

be the empirical process. The properties of the stochastic set function $\alpha_n(A) = \int_A d\alpha_n$ are investigated when A runs over a class of Borel sets of I^d. Let $\mathcal{A}$ be the set of Borel sets of I^d having d-times differentiable boundaries. Then a large deviation theorem and a law of iterated logarithm are proved for $\sup_{A \in \mathcal{A}} \alpha_n(A)$. A strong invariance principle (uniform over $\mathcal{A}$) is also formulated.

THREE THEOREMS OF MULTIVARIATE EMPIRICAL PROCESS

P. Révész

I. INTRODUCTION

Let $x_1,x_2,\ldots$ be a sequence of independent r.v.'s uniformly distributed over the unit cube I^d of the d-dimensional Euclidean Space R^d. Further let F_n be the empirical distribution function based on the sample $x_1,x_2,\ldots,x_n$ and let

$$\alpha_n(x) = n^{1/2}(F_n(x)-x_1x_2\ldots x_d) \quad (x=(x_1,x_2,\ldots,x_d)\in I^d) \text{ be}$$

the empirical process. The stochastic set function $\alpha_n(A)=\int_A d\alpha_n$ (A is a Borel set of I^d) will be called empirical measure.

The behaviour of the empirical process is frequently investigated; much less is known about the empirical measure if A runs over a wider class of the Borel sets of I^d. Most of the papers are studying the empirical measure over the convex sets or over a subclass of the convex sets. Dudley's paper [3] suggests the idea to study the empirical measure over sets having smooth boundaries. Paper [11] was devoted to develop this idea in case $d=2$. In this paper we intend to generalize the results of [11] for arbitrary d. The methods of proofs of the present paper mostly follow those of [11]. Hence the proofs will be omitted in the cases when they can be obtained repeating the methods of [11] and applying the new ideas formulated in Sections III.2 and III.3 of this paper.

In this Introduction we summarize those results (regarding mostly for empirical processes) what will be extended for empirical measures in Section II. For a detailed and excellent survey see [6].

I.1. A large deviation theorem. The following result was proved in 1961 by Kiefer:

THEOREM A. ([7]) *For each* $d=1,2,\ldots$ *and* $\varepsilon>0$ *there is a constant* $C=C(\varepsilon,d)$ *such that*

$$P\{\sup_{x\in I^d} |\alpha_n(x)| \geq z\} \leq Ce^{-(2-\varepsilon)z^2}.$$

I.2. Laws of the iterated logarithm. As a strengthening of many previous results in 1971 Zaremba (see also Wichura 1973) proved:

THEOREM B. ([15]) *Let* $I=I(d)$ *be the class of the intervals of* I^d. *Then*

$$\lim_{n\to\infty}\sup \sup_{A\in I} [(2\log\log n)^{1/2}\alpha_n(A)-(\lambda(A)(1-\lambda(A)))^{1/2}] = 0$$

w.p.1.

Here and in what follows λ is the Lebesgue measure.

A result closely related to the latter one was obtained by Philipp in 1973:

THEOREM C. ([9]) *Let* $E=E(d)$ *be the class of ellipsoids with axis parallel to the coordinate axis. Then*

$$\limsup_{n\to\infty} \sup_{A\in E} (2\log\log n)^{1/2}\alpha_n(A) = 1/2$$

w.p. 1.

Philipp also proved:

THEOREM D. ([9]) *Let* $C=C(2)$ *be the class of the convex sets of* I^2. *Then*

$$\limsup_{n\to\infty} \sup_{A\in C} (2\log\log n)^{1/2}\alpha_n(A) = 1/2$$

w.p. 1.

Investigating the behaviour of the empirical measure over the convex sets of I^d $(d \geq 3)$ mostly negative results were obtained ([13],[16]). It is known that in case $d > 4$ the law of iterated logarithm does not hold, the case $d=3$ is unknown.

A very new direction of these researches was proposed by Finkelstein in 1971 ([5]). She carried over the Strassen's idea [12] (developped for sums of i.i.d.r. v.'s) to the empirical process and proved:

THEOREM E. ([5]) *Let* K *be the set of functions* f *defined on* I^1 *for which*

(i) $f(0)=f(1)=0$,

(ii) f *is absolutely continuous with respect to* λ,

(iii) $\int (f')^2 \leq 1$.

Then the sequence $\{(2\log\log n)^{1/2}\alpha_n(x)\}$ *is relatively compact and the set of its limit points (in the sup norm) is* K.

This result was extended to the d-dimensional case by Wichura [15].

I.3. Strong approximation. First some notations:

N.I.3.1. Let A be a class of the Borel sets of I^d and for any $A\in A$ let $\xi(A)$ be a r.v. The stochastic set function $\xi(A)$ is called a

(i) Wiener measure (over A) if $\xi(A)\in N(0,\lambda^{1/2}(A))$ and $E\xi(A)\xi(B)=\lambda(A\cap B)$ $(A,B\in A)$,

(ii) Brownian measure (over A) if $\xi(A)\in N(0,(\lambda(A)(1-\lambda(A))^{1/2})$ and $E\xi(A)\xi(B) = =\lambda(A\cap B)-\lambda(A)\lambda(B)$ $(A,B\in A)$.

(if $W(A)$ is a Wiener measure then $B(A)=W(A)-\lambda(A)W(I^d)$ is a Brownian measure.)

N.I.3.2. For any $t > 0$ and $A\in A$ let $K(A;t)$ be a r.v. $K(A;t)$ is called a Kiefer measure if $K(A;t)\in N(0,(t\lambda(A)(1-\lambda(A)))^{1/2})$ and $EK(A;t_1)K(B;t_2) = = \min(t_1t_2)[\lambda(A\cap B)-\lambda(A)\lambda(B)]$ $(A,B\in A)$.

On the strong approximation of the empirical measure by Brownian resp. Kiefer measure the following is known:

THEOREM F. *One can construct a probability space* $\{\Omega, S, P\}$, *a sequence* $\{X_n\}$ *of independent r.v.'s uniformly distributed over* I^d, *a sequence* $\{B_n(A)\}$ *of Brownian measures and a Kiefer measure* $K(A;t)$ *(all of them are defined on* Ω*) such that*

$$\sup_{A\in\mathcal{A}} |\alpha_n(A)-B_n(A)| = O(n^{-1/2}\log n) \qquad w.p.\ 1,$$

$$\sup_{A\in\mathcal{A}} |n^{1/2}\alpha_n(A)-K(A;n)| = O(\log^2 n) \qquad w.p.\ 1$$

if $d=1$ *and* $\mathcal{A}$ *is the class of the intervals (see* [8]*);*

$$\sup_{A\in\mathcal{A}} |\alpha_n(A)-B_n(A)| = O\left(n^{-\frac{1}{2(d+1)}}(\log n)^{3/2}\right) \qquad w.p.\ 1,$$

$$\sup_{A\in\mathcal{A}} |n^{1/2}\alpha_n(A)-K(A;n)| = O\left(n^{-\frac{d+1}{2(d+2)}}\log^2 n\right) \qquad w.p.\ 1$$

for any $d=1,2,\ldots$ *if* $\mathcal{A}$ *is the class of the intervals (see* [2]*);*

$$\sup_{A\in\mathcal{A}} |\alpha_n(A)-B_n(A)| = O(n^{-1/2}\log^2 n) \qquad w.p.\ 1$$

if $d=2$ *and* $\mathcal{A}$ *is the class of the intervals (see* [14]*);*

$$\sup_{A\in\mathcal{A}} |\alpha_n(A)-B_n(A)| = O(n^{-1/19}) \qquad w.p.\ 1,$$

$$\sup_{A\in\mathcal{A}} |n^{1/2}\alpha_n(A)-K(A;n)| = O(n^{1/25}) \qquad w.p.\ 1$$

if $d=2$ *and* $\mathcal{A}$ *is the class of the Borel sets of* I^2 *having twice differentiable boundaries* (see [11]).

I.5. Acknowledgement. The author is indebted to Professors W. Philipp and G. Tusnády for their valuable remarks. A number of the ideas of this paper borned during conversations with them.

II. THE THREE THEOREMS

II.1. Notations.

N.2.1.1. Let $F_1=F_1(d,M,\rho)$ be the class of functions f of $(d-1)$ variables for which

$$D^p_{p_1,p_2,\ldots,p_{d-1}} f = D^p f = \frac{\partial^p f}{\partial^{p_1}x_2\partial^{p_2}x_2\ldots\partial^{p_{d-1}}x_{d-1}}$$

$(p_1+p_2+\ldots+p_{d-1}=p;\ p_i=0,1,2,\ldots,d-1;\ 0 \leq x_i \leq 1,\quad p=1,2,\ldots,d-1,\ i=1,2,\ldots,d-1)$
exists and

$$|D^p f| \leq M, \qquad |D^{d-1}f(x)-D^{d-1}f(y)| \leq M\|x-y\|^{\rho}$$

$(x,y\in I^d,\ M > 0,\ 0 < \rho \leq 1,\quad p=1,2,\ldots,d-1)$.

N.2.1.2. Let F be a class of Borel measurable functions of $d-1$ variables. Let A(F) be the class of Borel sets

$$A = A(f_1^{(1)},f_1^{(2)},f_2^{(1)},\ldots,f_d^{(1)},f_d^{(2)}) =$$

$$(2.1.1) \qquad = \bigcap_{i=1}^{d} \{x=(x_1,x_2,\ldots,x_d):\max(0,f_i^{(1)}(x_1,x_2,\ldots,x_{i-1},x_{i+1},\ldots,x_d)) <$$

$$< x_i < \min(1,f_i^{(2)}(x_1,x_2,\ldots,x_{i-1},x_{i+1},\ldots,x_d))\}$$

where $f_i^{(j)}\in$F $(i=1,2,\ldots,d;\ j=1,2)$. Especially let $A_1=A_1(d,M,\rho)=A(F_1(d,M,\rho))$.

N.2.1.3. Let $A_o=A_o(d,M,\rho,S)$ be the class of Borel sets A of I^d which can be represented in the form:

$$A = \bigcup_{i=1}^{S} A_i \qquad (A_i\in A_1)$$

The class A_o is closely related to the class $I(d,\alpha,M)$ $(\alpha > d-1)$ defined by Dudley ([4]). The exact relationship is unknown, however I do believe in the following:

CONJECTURE. For any $M,\alpha = d-1+\rho$ there exist an $M^*=M^*(d,\alpha,M)$ and a $S = S(\alpha,d,M)$ such that

$$I(d,\alpha,M) \subset A_o(d,M^*,\rho,S)$$

N.2.1.4. Let A be any class of Borel sets of I^d. Then for any $0 < \tau \leq 1/2$ let

$$A^* = A^*(\tau) = \{A : A\in A,\ (\lambda(A)(1-\lambda(A)))^{1/2} \leq \tau\}.$$

II.2. <u>A large deviation theorem.</u> We present the following analogue of Theorem A.:

THEOREM 1. *For each* $0 < \varepsilon,\ \varepsilon_1 < 1,\ M > 0,\ 0 < \rho < 1,\ d=1,2,\ldots$ *there is a constant* $K=K(\varepsilon,\varepsilon_1,d,\rho,M)$ *such that*

$$P\{\sup_{A\in A_o^*} |\alpha_n(A)| \geq \tau z\} \leq \tau^{-\tau^{-K}} e^{-z^2/(2+\varepsilon)}$$

provided that $z^2\tau^{-\frac{8(d-1)}{\rho}-4} \leq n^{1-\varepsilon_1}$ $\quad (0 < \tau \leq 1/2)$.

II.3. A law of the iterated logarithm. First some notations:

N.2.3.1. Let $\mathcal{A}$ be a class of Borel sets of I^d. Then define the distance $\Theta(\mathcal{A})=\Theta_{\mathcal{A}}(\mu,\nu)$ between two signed measures μ and ν (defined on the Borel sets of I^d) by

$$\Theta(\mathcal{A}) = \sup_{A\in\mathcal{A}} |\mu(A)-\nu(A)|.$$

N.2.3.2. Let $G=G(d)$ be the set of Borel measurable functions g of d variables for which

(i) $g(x_1,x_2,\ldots,x_{i-1}, 0, x_{i+1},\ldots,x_d) = g(1,1,\ldots,1) = 0,$

(ii) g is absolutely continuous with respect to λ,

(iii) $\int_{I^d} \left(\frac{\partial^d g}{\partial x_1 \partial x_2 \ldots \partial x_d}\right)^2 d\lambda \le 1$

N.2.3.3. Let K be the set of signed measures K_g defined on the Borel sets of I^d by:

$$K_g(A) = \int_A dg \qquad (g\in G).$$

Now our result runs as follows:

THEOREM 2. *The sequence* $\{(2\log\log n)^{-1/2}\,\alpha_n(A)\}$ *is relatively compact with respect to* $\Theta_1=\Theta(\mathcal{A}_1)$ *and the set of its limit points is* K.

Relatively compact with respect to the distance Θ_1 with limit points K (see N.2.1.2. and N.2.3.1.) means: there exists a set $\Omega_0 \subset \Omega$ (the basic set) with $P(\Omega_0)=1$ such that

(i) for any $\omega\in\Omega_0$ and $\kappa\in K$ there exists a sequence $n_1=n_1(\omega,\kappa) < n_2=n_2(\omega,\kappa)<\ldots$ such that

$$\Theta_1((2\log\log n_k)^{-1/2}\alpha_{n_k},\kappa) \to 0 \qquad \text{w.p.1.}$$

and

(ii) for any $\omega\in\Omega_0$ and sequence $\{n_k\}$ there exist a subsequence $n_{k_1}=n_{k_1}(\omega) < n_{k_2} = n_{k_2}(\omega) <\ldots$ and a $\kappa\in K$ such that

$$\Theta_1((2\log\log n_{k_j})^{-1/2}\alpha_{n_{j_k}},\kappa) \to 0 \qquad \text{w.p.1.}$$

Our Theorem 2 implies:

THEOREM 2*.

$$\limsup_{n\to\infty} \sup_{A\in \mathcal{A}_o} [(2\log\log n)^{-1/2}\alpha_n(A)-(\lambda(A)(1-\lambda(A)))^{1/2}] = 0$$

w.p. 1.

This latter result is a generalization of Theorems B and C.

II.4. Strong approximation. We will prove the following

THEOREM 3. *One can construct a probability space* $\{\Omega,S,P\}$, *a sequence* $\{X_n\}$ *of independent r.v.'s uniformly distributed over* I^d, *a sequence* $\{B_n(A)\}$ *of Brownian measures and a Kiefer measure* $K(A;t)$ *(all of them are defined on* Ω*) such that*

$$\sup_{A\in \mathcal{A}_o} |\alpha_n(A)-B_n(A)| = O(n^{-\frac{\rho}{12(d+1)}}) \qquad w.p.\ 1,$$

$$\sup_{A\in \mathcal{A}_o} |n^{1/2}\alpha_n(A)-K(A;n)| = O(n^{\frac{3(d+1)}{6(d+1)+\rho}}) \qquad w.p.1.$$

III. PROOFS

III.1. Notations.

N.3.1.1. $I_j(d,r)=\overset{d}{\underset{i=1}{\times}} [j_iR^{-1},(j_i+1)R^{-1})$ where $R=2^r$ $(j=j_1,j_2,\ldots,j_d)$; $j_i=0,1,2,\ldots,R-1$; $i=1,2,\ldots,d$; $d=1,2,\ldots$; $r=0,1,2,\ldots)$.

N.3.1.2. Let $F_3=F_3(d)$ be the class of rational functions f of $(d-1)$ variables with grad $f\le d-1$. That is

$$f = \sum_{\substack{p_1+p_2+\ldots+p_{d-1}\le d-1;\\ p_i=0,1,2,\ldots,d-1}} a(p_1,p_2,\ldots,p_{d-1})x_1^{p_1}x_2^{p_2}\ldots x_{d-1}^{p_{d-1}}\ .$$

N.3.1.3. Let $F_2=F_2(d,t,M)$ be the class of functions f of $(d-1)$ variables which can be represented in the form:

$$f(x) = f_j(x) \qquad \text{if} \qquad x\in I_j(d-1,t)$$

where $f_j(x)\in F_3(d)$ $(j=(j_1,j_2,\ldots,j_{d-1})$; $j_i=0,1,2,\ldots,T-1$; $T=2^t)$ and $|D^1f_j(x)|\le M$ if x is an inner point of $I_j(d-1,t)$.

N.3.1.4. Let $A_3=A_3(d)$ be the class of Borel sets A of I^d which can be represented either in the form:

$$A=A^{(1)}(f)=\{x=x_1,x_2,\ldots,x_d) : 0 < x_d < \min(1,f(x_1,x_2,\ldots,x_{d-1}))\},$$

or

$$A=A^{(2)}(f)=\{x=(x_1,x_2,\ldots,x_d) : \max(0,f(x_1,x_2,\ldots,x_{d-1})) < x_d < 1\}$$

where $f \in F_3$,

N.3.1.5. Let

$$A_2 = A_2(d,t,M) = A(F_2(d,t,M))$$

(see N.2.1.2 and N.3.1.3).

N.3.1.6. For any Borel set $A \subset I^d$ let

$$A(r) = \bigcup_{\{j:I_j(d,r) \subset A\}} I_j(d,r), \qquad \tilde{A}(r) = \bigcup_{\{j:I_j(d,r)\cap A \neq 0\}} I_j(d,r) - A(r),$$

$$\bar{A}(r) = A(r+1) - A(r).$$

N.3.1.7. Let A be a class of Borel sets of I^d. Then put

$$\mathsf{A}(r) = \{A(r) : A \in \mathsf{A}\},$$
$$\bar{\mathsf{A}}(r) = \{\bar{A}(r) : A \in \mathsf{A}\},$$
$$\tilde{\mathsf{A}}(r) = \{\tilde{A}(r) : A \in \mathsf{A}\},$$
$$\bar{\mathsf{A}}^*(r,\tau) = \bar{\mathsf{A}}^*(r) = \{\bar{A}(r) : A \in \mathsf{A}^*\},$$
$$\tilde{\mathsf{A}}^*(r,\tau) = \tilde{\mathsf{A}}^*(r) = \{\tilde{A}(r) : A \in \mathsf{A}^*\}$$

(see N.2.1.4 and N.3.1.6)

N.3.1.8. For any finite set B let card B be the number of the elements of B.

N.3.1.9. Let $\mathsf{A}_4 = \mathsf{A}_4(d,r)$ be the class of Borel sets A of I^d which can be represented in the form

$$A = \bigcup I_{j_\alpha}(d,r)$$

where j_α runs over an arbitrary subset of the lattice points: $j=(j_1,j_2,\ldots,j_d)$ $(j_i=0,1,2,\ldots,R)$.

N.3.1.10. Let $f \in F_1$ and define the functions ${}_{(t)}f$, by

$${}_{(t)}f(x) = f_j(x) \qquad \text{if} \qquad x \in I_j(d-1,t)$$

where

$$f_j = \sum a_j(p_1,p_2,\ldots,p_{d-1}) x_1^{p_2} x_2^{p_2} \ldots x_{d-1}^{p_{d-1}} \in F_3$$

$(j=(j_1,j_2,\ldots,j_{d-1}),\ j_i=0,1,2,\ldots,T-1,\ p_1+p_2+\ldots+p_{d-1} = p,\ p=0,1,2,\ldots,d-1)$ and the coefficients a_j are defined such that the equations

$$\frac{\partial^p f_j(y_1,y_2,\ldots,y_{d-1})}{(\partial y_1)^{p_1}(\partial y_2)^{p_2}\ldots(\partial y_{d-1})^{p_{d-1}}} = \frac{\partial^p f(y_1,y_2,\ldots,y_{d-1})}{(\partial y_1)^{p_1}(\partial y_2)^{p_2}\ldots(\partial y_{d-1})^{p_{d-1}}}$$

$$y_i = y_i(j) = \frac{j_i+(j_i+1)}{2T}$$

$(p=0,1,2,\ldots,d-1)$ should hold. Further let

$${}^{(2)}_{(t)}f(x) = {}_{(t)}f(x) - c_j^{(-)} \qquad \text{if} \qquad x\in I_j(d-1,t)$$

$${}^{(1)}_{(t)}f(x) = {}_{(t)}f(x) + c_j^{(+)} \qquad \text{if} \qquad x\in I_j(d-1,t)$$

where

$$c_j^{(-)} = \max\{0, \sup_{x\in I_j(d-1,t)} ({}_{(t)}f(x)-f(x))\}$$

$$c_j^{(+)} = \max\{0, \sup_{x\in I_j(d-1,t)} (f(x)-{}_{(t)}f(x))\}.$$

By words ${}_{(t)}f$ is an approximation of f by polynomials, ${}^{(1)}_{(t)}f$ (resp. ${}^{(2)}_{(t)}f$) are upper (resp. lower) estimations.

N.3.1.11. Let $A\in A_1$ be represented in the form (2.1.1) then let

$$A_*(t) = A({}^{(1)}_{(t)}f_1^{(1)}, {}^{(2)}_{(t)}f_1^{(2)}, \ldots, {}^{(1)}_{(t)}f_d^{(1)}, {}^{(2)}_{(t)}f_d^{(2)}),$$

$$A_*^+(t) = A_*(t+1) - A_*(t),$$

$$A_*^-(t) = A_*(t) - A_*(t+1),$$

$$A^*(t) = A({}^{(2)}_{(t)}f_1^{(1)}, {}^{(1)}_{(t)}f_1^{(2)}, {}^{(2)}_{(t)}f_2^{(1)}, {}^{(1)}_{(t)}f_2^{(2)}, \ldots, {}^{(2)}_{(t)}f_d^{(1)}, {}^{(1)}_{(t)}f_d^{(2)}),$$

$$\tilde{A}_*(t) = A^*(t) - A_*(t).$$

Clearly for any $t=0,1,2,\ldots$ we have

$$A_*(t) \subset A \subset A^*(t).$$

N.3.1.13. Let L be a subspace of the Euclidean Space R^d and denote P^d_L be the projection of R^d onto L.

III.2. Lemmas

LEMMA 1.

$$\text{card } A_3(r) \le 2(2(R+1)^d, d)^{d-1}$$

(see N.3.1.4. and N.3.1.7.)

PROOF. Let ψ_d be the number of the coefficients of the elements of F_3 and define the one-to-one map $M \in F_3 \leftrightarrow R^{\psi_d}$ as $Mf=a_f$ where the coordinates of the vector a_f are the coefficients of f in an arbitrary but fixed order. Further let $x_i=(x_{i1},x_{i2},\dots,x_{id})$ $(x_{ij}=0,1/R,2/R,\dots,(R-1)/R,\ i=1,2,\dots,(R+1)^d)$ be the sequence of the lattice points of I^d in a given order.

Consider the set $F_i \subset F_3$ of functions which are passing over the lattice point x_i, i.e. $f \in F_i$ if $x_{id}=f(x_{i1},x_{i2},\dots,x_{i,d-1})$. Clearly $G_i=MF_i$ $(i=1,2,\dots,(R+1)^d)$ is a hyperplane in R^{ψ_d}.

For each $a=a_f \in R^{\psi_d}$ define the sequence

$$T_a = \{y_1(x),z_1(a),y_2(a),z_2(a),\dots,y_{\psi_d}(a),z_{\psi_d}(a)\}$$

as follows:

$y_1(a)=i_1$ if i_1 is the smallest integer for which

$$\|a-G_{i_1}\| \le \|a-G_j\| \qquad (j=1,2,\dots,(R+1)^d),$$

$$z_1(a) = \begin{cases} 0 & \text{if } f(x_{i_1 1},x_{i_1 2},\dots,x_{i_1,d-1}) \ge x_{i_1 d}, \\ 1 & \text{if } f(x_{i_1,1},x_{i_1 2},\dots,x_{i_1,d-1}) < x_{i_1,d}. \end{cases}$$

Further let $y_2(a)=i_2$ if i_2 is the smallest integer $(\ne i_1)$ for which

$$\|P^{\psi_d}_{G_{i_1}} a-G_{i_1}\cap G_{i_2}\| \le \|P^{\psi_d}_{G_{i_1}} a-G_{i_1}\cap G_j\| \qquad (j\ne i_1)$$

and let

$$z_2(a) = \begin{cases} 0 & \text{if } f(x_{i_2 1},x_{i_2 2},\dots,x_{i_2 d-1}) \ge x_{i_2 d}, \\ 1 & \text{if } f(x_{i_2 1},x_{i_2 2},\dots,x_{i_2 d-1}) < x_{i_2 d}. \end{cases}$$

If $y_\ell(a)$, $z_\ell(a)$ are defined then let $y_{\ell+1}(a)=i_{\ell+1}$ if $i_{\ell+1}$ is the smallest integer (different from $i_1,i_2,\dots,i_\ell$) for which

$$\|P^{\psi_d}_{G_{i_\ell}} P^{\psi_d}_{G_{i_{\ell-1}}} \dots P^{\psi_d}_{G_{i_1}} a-G_{i_1}\cap G_{i_2}\cap\dots\cap G_{i_{\ell+1}}\| \le$$

$$\le \|P^{\psi_d}_{G_{i_\ell}} P^{\psi_d}_{G_{i_{\ell-1}}} \dots P^{\psi_d}_{G_{i_1}} a-G_{i_1}\cap G_{i_2}\cap\dots\cap G_{i_\ell}\cap G_j\|$$

where j is different from the integers $i_1,i_2,\dots,i_\ell$. Let $z_{i_{\ell+1}}(a)$ be defined by

$$z_{i_{\ell+1}}(a) = \begin{cases} 0 & \text{if } f(x_{i_{\ell+1} 1},x_{i_{\ell+1} 2},\dots,x_{i_{\ell+1} d-1}) \ge x_{i_{\ell+1} d}, \\ 1 & \text{if } f(x_{i_{\ell+1} 1},x_{i_{\ell+1} 2},\dots,x_{i_{\ell+1} d-1}) < x_{i_{\ell+1} d}. \end{cases}$$

Since $G_{i_1}\cap G_{i_2}\cap\dots\cap G_{i_\ell}$ is a $(\psi_d-\ell)$ dimensional Euclidean Space, the procedure will be finished after ψ_d^ℓ steps.

It is easy to check that $(A^{(1)}(f_1))(r)=(A^{(1)}(f_2))(r)$ if $T_{a_{f_1}}=T_{a_{f_2}}$. (See N.3.1.4 and N.3.1.6). Since $\psi_d \le d^{d-1}$ our statement follows.

This lemma implies:

LEMMA 2.

$$\text{card } \mathsf{A}_2^*(r) \le \text{card } \mathsf{A}_2(r) \le (2(R+1)^d)^{d^{d-1}2Td} \le R^{4Td^{d+1}},$$

$$\text{card } \bar{\mathsf{A}}_2^*(r) \le \text{card } \bar{\mathsf{A}}_2(r) \le (2(2R+1))^{d^{d-1}2Td} \le R^{4Td^{d+1}},$$

$$\text{card } \widetilde{\mathsf{A}}_2^*(r) \le \text{card } \widetilde{\mathsf{A}}_2(r) \le (2(R+1)^d)^{d^{d-1}2Td} \le R^{4Td^{d+1}}$$

if $R \ge 8$, $d \ge 2$. (see N.3.1.5. and N.3.1.7.)

The next Lemma is straight-forward by N.3.1.5. and N.3.1.6.

LEMMA 3. *For every* $A \in \mathsf{A}_2$ *and* $r \ge t$ *we have*

$$\lambda(\bar{A}(r)) \le 2Md^{3/2}R^{-1}, \quad \lambda(\widetilde{A}(r)) \le 2Md^{3/2}R^{-1}.$$

LEMMA 4. *Let* $f \in \mathsf{F}_1$, $A \in \mathsf{A}_1$ *then*

$$\sup_{x \in I^d} |f(x) - {}_{(t)}f(x)| \le \frac{(d-1)^{d-1}}{(d-1)!} \left(\frac{1}{2T}\right)^{d-1} \left(\frac{\sqrt{d}}{2T}\right)^{\rho} M \le$$

$$\le 2^d M T^{-d-\rho+1},$$

$$\lambda(A_*^+(t)) \le 2^{d+1} d M T^{-d-\rho+1},$$

$$\lambda(A_*^-(t)) \le 2^{d+1} d\, M T^{-d-\rho+1}$$

$$\lambda(\widetilde{A}_*(t)) \le 2^{d+2} d\, M T^{-d-\rho+1}.$$

where $T=2^t$. (see N.2.1.2 and N.3.1.11.).

PROOF. This Lemma follows via the Taylor expansion of f.

Applying again the Taylor expansion for $D^1 f$ one gets:

LEMMA 5. *Let* $A \in \mathsf{A}_1(d,M,\rho)$ *and let* t *be large enough, say* $t \ge t_o(d,M,\rho)$. *Then*

$$A_*(t) \in \mathsf{A}_2(d,t,M+1),$$
$$A^*(t) \in \mathsf{A}_2(d,t,M+1)$$

and $A_*^+(t)$, $A_*^-(t)$, $\widetilde{A}_*(t)$ *are the sums of* $2d$ *elements of* $\mathsf{A}_2(d,t,M+1)$.

III.3. Proof of Theorem 1. For sake of simplicity the proof will be given only in the case $s=1$, that is only for the class A_1 instead of A_o. The general case can be treated similarly.

In this proof there will be used frequently the

BERNSTEIN INEQUALITY (see e.g. [10] p. 387-389). *For any Borel set* $A \subset I^d$ *and*

$0 < \varepsilon < 1$ *we have*

$$P\{|\alpha_n(A)| \geq z(\lambda(A)(1-\lambda(A)))^{1/2}\} \leq 2e^{-z^2/(2+\varepsilon)}$$

provided that $0 < z < \varepsilon/3(n\lambda(A)(1-\lambda(A)))^{1/2}$.

LEMMA 6. *For any* $0 < \varepsilon < 1$ *and* $d=1,2,\ldots$ *we have*

$$P\{\sup_{A\in \mathcal{A}_2} |\alpha_n(\bar{A}(r))| \geq (2Md^{3/2}R^{-1})^{1/2}y_r\} \leq 2R^{4Td^{d+1}}e^{-y_r^2/(2+\varepsilon)} \tag{3.1}$$

provided that $0 < y_r < \varepsilon/3(Md^{3/2}nR^{-1})^{1/2}$.

PROOF. This Lemma is a straight consequence of Lemmas 2., 3. and the Bernstein inequality.

LEMMA 7. *For any* $0 < \varepsilon < 1$, $q > 3$ *and* $d=1,2,\ldots$ *we have*

$$P\{n^{1/2} \sup_{A\in \mathcal{A}_2} \int_{\tilde{A}(r)} dF_n \geq 2q\varepsilon^{-1}(2Md^{3/2}R^{-1})^{1/2}y_r\} \leq \tag{3.2}$$

$$\leq 2R^{4Td^{d+1}}e^{-y_r^2/(2+\varepsilon)}$$

provided that $\varepsilon q^{-1}(2Md^{3/2}nR^{-1})^{1/2} \leq y_r \leq \varepsilon/3(2Md^{3/2}nR^{-1})^{1/2}$.

PROOF. Since (by Lemma 3.)

$$n^{1/2} \sup \lambda(\tilde{A}(r)) \leq 2Md^{3/2}n^{1/2}R^{-1} =$$

$$= (q\varepsilon^{-1}(2Md^{3/2}R^{-1})^{1/2})(q^{-1}\varepsilon(2dnR^{-1})^{1/2}) \leq$$

$$\leq q\varepsilon^{-1}(2Md^{3/2}R^{-1})^{1/2}y_r$$

we have (by Lemmas 2., 3. and the Bernstein inequality

$$P\{n^{1/2} \sup \int_{\tilde{A}(r)} dF_n \geq 2q\varepsilon^{-1}(2Md^{3/2}R^{-1})^{1/2}y_r\} \leq$$

$$\leq P\{\sup|\alpha_n(\tilde{A}(r))| \geq q\varepsilon^{-1}(2Md^{3/2}R^{-1})^{1/2}y_r\} \leq$$

$$\leq P\{\sup|\alpha_n(\tilde{A}(r))| \geq (2Md^{3/2}R^{-1})^{1/2}y_r\} \leq$$

$$\leq 2R^{4Td^{d+1}}e^{-y_r^2/(2+\varepsilon)}.$$

LEMMA 8. *For any* $0 < \varepsilon,\ \varepsilon_1 < 1$, $M > 0$ *and* $d=1,2,\ldots$ *there exists a positive constant* $C=C(\varepsilon,\varepsilon_1,d,M)$ *such that*

(3.3) $$P\{\sup_{A\in A_2} |\alpha_n(A)-\alpha_n(A(r))| \geq CzR^{-1/2}\} \leq CR^{4Td^{d+1}} e^{-z^2/(2+\varepsilon)}$$

$(r=3,4,\ldots)$ *provided that* $z^2R \leq n^{1-\varepsilon_1}$.

PROOF. Statement (3.3) automatically holds true if $z < Q=(12Td^{d+1})^{1/2}$ (and $r \geq 3$), or if n is less than any fixed constant n_o (say). Hence we can assume that $z \geq Q$ and $n \geq n_o$.

Clearly we have

$$\sup|\alpha_n(A)-\alpha_n(A(r))| \leq \sum_{j=r}^{\ell-1} \sup|\alpha_n(\bar{A}(j))| + n^{1/2}\sup \int_{\widetilde{A}(\ell)} dF_n + n^{1/2}\sup\lambda(\widetilde{A}(\ell))$$

$$(\ell=r,r+1,\ldots).$$

Put

$$y_j = (Q^2(j-r)+z^2)^{1/2}, \quad j=r,r+1,\ldots$$

and let ℓ be the largest integer for which

$$2^\ell y_\ell^2 \leq \frac{\varepsilon^2}{9}(2Md^{3/2}n).$$

Our conditions clearly define uniquely such an ℓ and this ℓ will satisfy the following relations too:

(i) $y_j \leq \varepsilon/3(2Md^{3/2}n2^{-j})^{1/2}$ if $j=r,r+1,\ldots,\ell,$

(ii) $y_\ell \geq \varepsilon q^{-1}(2Md^{3/2}n2^{-\ell})$ if $q \geq 5$

(iii) $$n^{1/2}\lambda(\widetilde{A}(\ell)) \leq n^{1/2}2Md^{3/2}2^{-\ell} \leq \frac{18(Q^2(\ell-r)+z^2)}{\varepsilon^2 2Md^{3/2}n} n^{1/2}2Md^{3/2} \leq \frac{z^2}{n^{1/2}} \frac{18(\ell-r+1)}{\varepsilon^2} \leq$$

$$\leq zR^{-1/2}n^{\varepsilon_1/2} \frac{18(\ell-r+1)}{\varepsilon^2} \leq zR^{-1/2}$$

if n is big enough.

The above inequalities (i), (ii) show that Lemmas 6. and 7. are applicable. Using these Lemmas and (iii) we get:

$$P\{\sup|\alpha_n(A)-\alpha_n(A(r))| \geq zR^{-1/2}+\sum_{j=r}^{\ell-1}\left(2Md^{3/2}2^{-j}\right)^{1/2}y_j+2q\varepsilon^{-1}\left(2Md^{3/2}2^{-\ell}\right)^{1/2}y_\ell\} \leq$$

$$\leq \sum_{j=r}^{\ell} 2.2^{j4Td^{d+1}} e^{-y_j^2/(2+\varepsilon)} \leq CR^{4Td^{d+1}} e^{-z^2/(2+\varepsilon)} .$$

Now (3.3) follows from the following inequality:

$$zR^{-1/2}+\sum_{j=r}^{\ell-1}(2Md^{3/2}2^{-j})^{1/2}y_j+2q\varepsilon^{-1}(2Md^{3/2}2^{-\ell})^{1/2}y_\ell \le$$

$$\le zR^{-1/2}\{1+(2Md^{3/2})^{1/2}\sum_{j=0}^{\infty}2^{-j/2}+2q\varepsilon^{-1}(2Md^{3/2})^{1/2}2^{-\frac{\ell-r}{2}}\}+$$

$$+R^{-1/2}\{Q(2Md^{3/2})^{1/2}\sum_{j=0}^{\infty}(j2^{-j})^{1/2}+2q\varepsilon^{-1}Q(2Md^{3/2})^{1/2}((\ell-r)2^{-(\ell-r)})^{1/2}\}\le$$

$$\le C\,z\,R^{-1/2}.$$

LEMMA 9. *For any* $0<\varepsilon,\ \varepsilon_1<1$ *and* $d=1,2,\dots$ *there exists a positive constant* $C_1=C(\varepsilon,\varepsilon_1,d,M)$ *such that*

$$P\{\sup_{A\in\mathcal{A}_2^*}|\alpha_n(A)|\ge z\tau\}\le(C_1\tau^{-2})^{4Td^{d+1}}e^{-z^2/(2+\varepsilon)} \tag{3.4}$$

provided that $z^2\tau^{-2}\le n^{1-\varepsilon}$.

PROOF. Clearly we have

$$\sup|\alpha_n(A)|\le\sup|\alpha_n(A(r))|+\sup|\alpha_n(A)-\alpha_n(A(r))|.$$

Let $0<\varepsilon_2<1$ and let r be the smallest integer for which $2^r\ge(2C(\varepsilon_2\tau)^{-1})^2$ where $C=C(\varepsilon,\varepsilon_1/2,d,M)$ is the constant of Lemma 8.

Bernstein inequality and Lemma 2. imply

$$P\{\sup|\alpha_n(A(r))|\ge(1-\varepsilon_2)z\tau\}\le 2R^{4Td^{d+1}}\exp\{-\frac{z^2(1-\varepsilon_2)^2}{2+\varepsilon}\}\le$$
$$\le 2(C^2\varepsilon_2^{-2}\tau^{-2})^{4Td^{d+1}}\exp\{-\frac{z^2(1-\varepsilon_2)^2}{2+\varepsilon}\} \tag{3.5}$$

and by Lemma 8.

$$P\{\sup|\alpha_n(A)-\alpha_n(A(r))|\ge\varepsilon_2 z\tau\}\le P\{\sup|\alpha_n(A)-\alpha_n(A(r))|\ge CzR^{-1/2}\}\le$$
$$\le CR^{4Td^{d+1}}e^{-z^2/(2+\varepsilon)}\le(C_1\tau^{-2})^{4Td^{d+1}}e^{-z^2/(2+\varepsilon)}. \tag{3.6}$$

By an appropriate choosing of ε_2, (3.5) and (3.6) imply (3.4).

LEMMA 10. *For any* $0<\varepsilon,\ \varepsilon_1<1,\ M>0$ *and* $d=1,2,\dots$ *there exists a positive constant* $C=C(\varepsilon,\varepsilon_1,d,M)$ *such that*

$$P\{\sup_{A\in\mathcal{A}_1}|\alpha_n(A_*^+(t))|\geq C(T^{-d-\rho+1})^{1/2}y_t\}\leq \tag{3.7}$$
$$\leq CT^{d+\rho-1})^{4Td^{d+1}}e^{-y_t^2/(2+\varepsilon)}$$

($T=2^t$) provided that $y_t^2T^{d+\rho-1}\leq n^{1-\varepsilon_1}$. (3.7) *holds true if* $A_*^+(t)$ *is replaced by* $A_*^-(t)$ *or* $\widetilde{A}_*(t)$.

PROOF. It is a simple consequence of Lemmas 4. and 9.

LEMMA 11. *For any* $0<\varepsilon,\ \varepsilon_1<1,\ M>0$ *and* $d=1,2,\ldots$ *there exists a positive constant* $C=C(\varepsilon,\varepsilon_1,d,M)$ *such that*

$$P\{\sup_{A\in\mathcal{A}_1} n^{1/2}\int_{\widetilde{A}_*(t)}dF_n\geq C(T^{-d-\rho/2+1})^{1/2}y_t\}\leq \tag{3.8}$$
$$\leq (CT^{d+\rho-1})^{4Td^{d+1}}e^{-y_t^2/(2+\varepsilon)}$$

provided that

$$y_t^2\,T^{d+(3/2)\rho-1}\geq n2^{d+2}dM$$

and

$$y_t^2T^{d-1+\rho}\geq n^{1-\varepsilon_1}.$$

PROOF. Since (by Lemma 4.)

$$n^{1/2}\lambda(\widetilde{A}_*(t))\leq n^{1/2}2^{d+2}dMT^{-d-\rho+1}=$$
$$=n^{1/2}(2^{d+2}dMT^{-d+1-(3/2)\rho})^{1/2}(2^{d+2}dMT^{-d+1-\rho/2})^{1/2}\leq \tag{3.9}$$
$$\leq(2^{d+2}dMT^{-d+1-\rho/2})^{1/2}\cdot y_t$$

we have

$$P\{\sup n^{1/2}\int_{\widetilde{A}_*(t)}dF_n\geq 2(2^{d+2}dMT^{-d+1-\rho/2})^{1/2}\,y_t\}\leq$$

$$\leq P\{\sup|\alpha_n(\widetilde{A}_*(t))|\leq(2^{d+2}dMT^{-d+1-\rho/2})^{1/2}\,y_t\}\leq$$

$$\leq P\{\sup|\alpha_n(\widetilde{A}_*(t))|\geq(2^{d+2}dMT^{-d+1-\rho})^{1/2}y_t\}$$

Now (3.8) follows from Lemmas 4., 5. and 9.

LEMMA 12. *For any* $0 < \varepsilon, \varepsilon_1 < 1$, $M > 0$ *and* $d=1,2,\ldots$ *there exists a positive constant* $C=C(\varepsilon,\varepsilon_1,M,d)$ *such that*

$$P\{\sup_{A\in\mathcal{A}_1} |\alpha_n(A_*(t))| \geq CzT^{-1/2(d+\rho-1)}\} \leq T^{T^C} e^{-z^2/(2+\varepsilon)} \tag{3.10}$$

provided that $z^2T^{d+\rho-1} < n^{1-\varepsilon_1}$.

PROOF. Clearly we have

$$\sup|\alpha_n(A)-\alpha_n(A_*(t))| \leq \sum_{k=t}^{s} \sup|\alpha_n(A_*^+(k))| +$$

$$+\sum_{k=t}^{s} \sup|\alpha_n(A_*^-(k))| + n^{1/2}\sup \lambda(\widetilde{A}_*(s))+n^{1/2}\sup\int_{\widetilde{A}_*(s)} dF_n.$$

Set

$$y_k = (2^{Q(k-t)} + z^2)^{1/2}, \qquad Q=d-1 \quad (k=t,t+1,\ldots)$$

and choose the integer s such a way that the inequalities

$$y_s^2\, 2^{s(d+\rho-1)} \leq n^{1-\varepsilon_2}$$

and

$$y_s^2\, 2^{s(d+(3/2)\rho-1)} \leq n\, 2^{d+2}dM$$

should hold. Clearly if ε_2 is small enough one can find such an s.

Applying (3.9) for s instead of t one gets

$$n^{1/2}\lambda(\widetilde{A}_*(s)) \leq Cz\, 2^{s\rho/4}. \tag{3.10}$$

Now Lemmas 4., 10., and 11. and the inequality (3.10) imply

$$P\{ \sup|\alpha_n(A)-\alpha_n(A_*(t))| \geq C \sum_{k=t}^{s} 2^{-k/2(d+\rho-1)} y_k\} \leq$$

$$\leq \sum_{k=t}^{s} (C2^{k(d+\rho-1)})^{2^k 4d^{d+2}} e^{-y_k^2/(2+\varepsilon)} \leq T^{T^K} e^{-z^2/(2+\varepsilon)}$$

if K is big enough.

This inequality and the simple inequality

$$\sum_{k=t}^{s} 2^{-(k/2)(d+\rho-1)} y_k \leq Cz\, 2^{t\rho/4}$$

prove (3.10).

PROOF OF THEOREM 1. Clearly we have

$$\sup|\alpha_n(A)| \le \sup|\alpha_n(A)-\alpha_n(A_*(t)| + \sup|\alpha_n(A_*(t))|.$$

Set $t=[\log_2(c\varepsilon_2^{-1}\tau^{-1})^{4/\rho}]$ where c is the constant of Lemma 11. Then by Lemmas 8. and 11.

$$P\{\sup|\alpha_n(A_*(t))| \ge (1-\varepsilon_2)z\tau\} \le (C_1\tau^{-2})^{4dT^{d+1}} \exp\{-\frac{(1-\varepsilon_2)^2 z^2}{(2+\varepsilon)}\}$$

and

$$P\{\sup|\alpha_n(A)-\alpha_n(A_*(t))| \ge \varepsilon_2 z\tau\} =$$

$$= P\{\sup|\alpha_n(A)-\alpha_n(A_*(t))| \ge CT^{-\rho/4}\frac{\varepsilon_2 z\tau}{C}T^{\rho/4}\} \le$$

$$\le P\{\sup|\alpha_n(A)-\alpha_n(A_*(t))| \ge CT^{-\rho/4}z\} \le$$

$$\le (C(C\varepsilon_2^{-1}\tau^{-1})^{4/\rho[2(d-1)+\rho]})^{4d^{d+1}+1}e^{-z^2/(2+\varepsilon)}.$$

what prvoes Theorem 1.

III.4. Proof of Theorem 2. Having our Theorem 1 and applying the method of Finkelstein ([5]) and that of Csâki ([1]), the proof is very easy.

LEMMA 13. *The sequence* $\{(2\log\log n)^{-1/2}\alpha_n(A)\}$ *is relatively compact with respect to* $\Theta_4=\Theta(A_4)$ *and the set of its limit points is* K. (See N.2.3.1. and N.3.1.9)

This Lemma can be obtained on the same way as Lemma 4 of [5] was obtained. It is also a consequence of the results of Wichura [15].

LEMMA 14. *For any* $0<\varepsilon<1$, $t=1,2,\dots$, $d=1,2,\dots$ *and* $0<\varepsilon_1<1$ *there exist positive constants* $C_1=C_1(\varepsilon,\varepsilon_1,d)$ *and* $C_2=C_2(\varepsilon,\varepsilon_1,d,R)$ *such that*

$$P\{\sup_{2^k<n\le 2^{k+1}}\ \sup_{A\in A_2}|\alpha_n(A)-\alpha_n(A(r))| \ge C_1 zR^{-1/2}\} \le C_2 e^{-z^2/(2+\varepsilon)}$$

$(k=1,2,\dots)$ *provided that* $z^2R \le 2^{k(1-\varepsilon_1)}$

PROOF. Let $\xi_n=n\sup(\alpha_n(A)-\alpha_n(A(r)))=n\sup\alpha_n(A-A(r))$. Then $E(\xi_{n+1}|\xi_1,\dots,\xi_n)\ge \xi_n$ (w.p.1) that is $\{\xi_n\}$, as well as $\{e^{t\xi_n^2}\}$ $(t>0)$ are semimartingales. Now applying the semimartingale inequality and Lemma 8 we get Lemma 14.

LEMMA 15. *For any* $0<\varepsilon,\ \varepsilon_1<1$, $d=1,2,\dots;$ $t=1,2,\dots$ *and* $M>0$ *there exist positive constants* $C_1=C_1(\varepsilon,\varepsilon_1,M,d)$ *and* $C_2=C_2(\varepsilon,\varepsilon_1,M,d,t)$ *such that*

$$P\{\sup_{2^k<n\le 2^{k+1}} \sup_{A\in A_1} |\alpha_n(A)-\alpha_n(A_*(t))| \ge C_1 z T^{-\rho/4}\} \le C_2 e^{-z^2/2+\varepsilon)}$$

provided that $z^2T^{2(d-1)+\rho} \le n^{1-\varepsilon_1}$.

This Lemma can be proved on the same way as Lemma 14.

LEMMA 16. *For any* $r=1,2,\ldots$ *there exists a set* $\Omega_0^{(r)} \subset \Omega$ *such that* $P(\Omega_0^{(r)})=1$ *and for any* $\omega\in\Omega_0^{(r)}$ *among the inequalities*

$$\sup_{A\in A_2} (2\ \mathrm{loglog}\ n)^{-1/2}|\alpha_n(A)-\alpha_n(A(r))| \ge C R^{-1/2},$$

$$\sup_{A\in A_1} (2\mathrm{loglog}\ n)^{-1/2}|\alpha_n(A_*(t))| \ge CT^{-\rho/4}$$

only finitely many will be occured where the constant C *depends on* d *and* M *only.*

PROOF. This Lemma is a straight consequence of Lemmas 14., 15. and the Borel-Cantelli lemma.

LEMMA 17. *For any* $\varepsilon > 0$ *there exists an* $r_o=r_o(\varepsilon)$ *such that*

$$\sup_{K\in K} \sup_{A\in A_2} |K(A-A(r))| \le \varepsilon$$

and

$$\sup_{K\in K} \sup_{A\in A_1} |K(A-A_*(r))| \le \varepsilon$$

if $r \ge r_o$ (see N.2.3.2., N.2.1.2., N.3.1.5., N.3.1.6. and N.3.1.11.

PROOF. This simply follows from the Cauchy-inequality.

LEMMA 18. *The sequence* $\{(2\mathrm{loglog}\ n)^{-1/2}\alpha_n(A)\}$ *is relatively compact with respect to* $\Theta_2=\Theta(A_2)$ *and the set of its limit points is* K. (See N.3.1.5. and N.2.3.1.).

PROOF. This follows from Lemmas 13., 15. and 17.

Now our Theorem 2 follows from Lemmas 16., 17. and 18.

III.5. Proof of Theorem 3. This proof is a simple copying of the proof of Theorem 2 of [11]. Here we only sketch the main steps.

Step.1. Construct a sequence $\{B_n(x), x\in I^d\}$

$$\sup|B_n(j_1R_n^{-1},j_2R_n^{-1},\ldots,j_dR_n^{-1})-\alpha_n(j_1R_n^{-1},j_2R_n^{-1},\ldots,j_dR_n^{-1})| = O(n^{-\frac{1-d\alpha}{2}}\log^2 n) \quad (3.10)$$

$(j_i = 0,1,2,\dots,R_n;\ R_n = 2^{r_n} = O(n^{\alpha})\ \ (0 < \alpha < d^{-1})$ w.p.1.

(For the details see [11] or [2].)

Step 2. Define the Brownian measures of the sets $A \in \mathcal{A}_2$ by $B_n(A) = \lim_{r \to \infty} B_n(A(r))$ where $B_n(A(r))$ can be defined in the natural way using the processes $B_n(x)$ constructed in Step I. One can prove that the convergence of $B_n(A(r))$ to $B_n(A)$ is uniform in $A \in \mathcal{A}_2$ w.p.1:

$$(3.11)\qquad \sup_{A \in \mathcal{A}_2} |B_n(A) - B_n(A(r))| = O(R^{-1/2}(\log R)^{1/2})$$

w.p.1. (see Lemma 2 of [11]). Using (3.10), (3.11) and our Lemma 7 we get

$$(3.12)\qquad \sup_{A \in \mathcal{A}_2} |B_n(A) - \alpha_n(A)| = O(n^{-\frac{1}{2(d+1)}}(\log n)^{3/2})$$

w.p.1.

Step 3. Define the Brownian measures of the sets $A \in \mathcal{A}_1$ by $B_n(A) = \lim_{t \to \infty} B_n(A_*(t))$ where $B_n(A_*(t))$ was defined in Step 2. One can prove that the convergence of $B_n(A_*(t))$ to $B_n(A)$ is uniform in $A \in \mathcal{A}_1$ w.p.1:

$$(3.13)\qquad \sup_{A \in \mathcal{A}_1} |B_n(A) - B_n(A_*(t))| = O(T^{-\rho/4})$$

w.p.1. (see Lemma 5* of [11] and Lemma 12). Using (3.12), (3.13) and our Lemma 11 we get (2.1).

Step 4. In order to get (2.2) from (2.1) we can straightly follow the method of [2] or [11].

REFERENCES

[1] CSÁKI, E. (1968). An iterated logarithm law for semimartingales and its application to empirical distribution function. *Studia Sci. Math. Hung.* 3 287-292.

[2] CSÖRGŐ, M. and RÉVÉSZ, P. (1975). A new method to prove Strassen type laws of invariance principle. II. *Z. Wahrscheinlichkeitstheorie und Verw. Gebiete* 31 261-269.

[3] DUDLEY, R.M. (1973). Sample function of Gaussian process. *Ann. Probability* 1 66-103.

[4] DUDLEY, R.M. (1974). Metric entropy of some classes of sets with differentiable boundaries. *Journal of Approximation Theory* 10 227-236.

[5] FINKELSTEIN, H. (1971). The law of iterated logarithm for empirical distributions. *Ann. Math. Statist.* 42 607-615.

[6] GAENSSLER, P. and STUTE, W. (1975). A survey on some results for multidimensional empirical processes in the I.I.D. case. Technical report. Ruhr-Universitäet Bochum.

[7] KIEFER, J. (1961). On large deviations of the empiric d.f. of vector chance variables and a law of the iterated logarithm. *Pacific J. Math.* 11 649-660.

[8] KOMLÓS, J., MAJOR, P. and TUSNÁDY, G. (1975). An approximation of partial sums of independent r.v.'s and the sample d.f.I. *Z. Wahrscheinlichkeitstheorie und Verw. Gebiete* 32 111-131.

[9] PHILIPP, W. (1973). Empirical distribution functions and uniform distribution mod 1. *Diophantine approximation and its applications*. Academic Press, New York.

[10] RÉNYI, A. (1970). *Probability Theory*. Akadémiai Kiadó, Budapest.

[11] RÉVÉSZ, P. (1976). On strong approximation of the multidimensional empirical process. To appear. *Ann. Probability*.

[12] STRASSEN, V. (1964). An invariance principle for the law of the iterated lograithm. *Z. Wahrscheinlichkeitstheorie und Verw. Gebiete* 3 211-226.

[13] STUTE, W. (1975). Convergence rates for the isotrope discrepancy. Technical report, Ruhr-Universität Bochum.

[14] TUSNÁDY, G. (1976). A remark on the approximation of the sample DF in the multidimensional case. To appear. *Periodica Math. Hung.*

[15] WICHURA, M.J. (1973). Some Strassen-type laws of the iterated logarithm for multiparameter stochastic processes with independent increments. *Ann. Probability* 1 272-296.

[16] ZAREMBA, S.K. (1970). La discrépance isotrope et l'integration numérique. *Ann. Mat. Pura Appl. (IV)* 87, 125-135.

[17] ZAREMBA, S.K. (1971). Sur la discrépance des suites aléatoires. *Z. Wahrscheinlichkeitstheorie und Verw. Gebiete* 20 236-248.

WEAK CONVERGENCE TO STABLE LAWS
BY MEANS OF A WEAK INVARIANCE PRINCIPLE

by

Gordon Simons[1]

and

William Stout[2]

Department of Mathematics
University of Illinois
Urbana, Illinois 61801

1. Introduction.

This paper presents a new method for establishing convergence in law of normed sums of independent identically distributed (iid) random variables. A distincitive feature of this method is that no use of transforms is made (except the asymptotic distributional form of stable laws is taken for granted). The method is probabilistic. It depends completely on the establishment of an appropriate invariance principle. The invariance principle introduced here is analogous to the almost sure invariance principles appearing in the literature (cf., Strassen (1964, 1965), Csörgö and Révész (1975), Kolmós, Major and Tusnády (1975), Philipp and Stout (1975)), except almost sure convergence is replaced by convergence in probability or, what is equivalent in this setting, convergence in law. For this reason, the terminology "weak invariance principle" is used. The weak invariance principle encapsulated in Theorem 1 below is based upon very elementary notions of probability. This contrasts sharply with the typical almost sure invariance principle, which depends on the existence and properties of Brownian motion and frequently on some form of Skorokhod embedding. Our weak invariance principle is designed to demonstrate the sufficiency of the classical assumptions associated with the weak convergence of normed sums to stable laws. Perhaps more sophisticated weak invariance principles can be found which would apply more widely within the scope of the central limit problem.

[1] The work of this author was supported by the United States Air Force Office of Scientific Research under Grant No. AFOSR-75-2796.

[2] The work of this author was supported by the United States National Science Foundation.

Our main result, Theorem 1, is concerned with random variables which are within the domain of partial attraction of some infinitely divisible law (implied by assumption A2 below) but outside of the domain of partial attraction of a normal law (i.e., satisfying assumption A1). The latter restriction is essential to our approach. It implies that our theory does not concern itself with the convergence of normed sums to the normal law. But it does apply to all other stable laws. In this regard, it should be pointed that Root and Rubin (1973) have already established the normal convergence criterion by probabilistic methods. The effect of our work is to bring probabilistic methods to another area of the general central limit problem.

By-products of our research are two alternative characterizations (see A1" and A1‴) of the class of distributions which are within the domain of partial attraction of a normal law. We believe these are new. They may be of independent interest. We also obtain a result concerning domains of partial attraction.

In Section 2 a statement of our main result and subsequent corollaries is given. A proof of this result will appear elsewhere. However, in order that the method of proof can be understood, a special case of Theorem 1 is stated and proved in Section 3. The proof of Theorem 1 is quite similar, except that certain properties of slowly varying functions play a crucial role.

2. Statement of the results.

The weak invariance principle described in Theorem 1 basically says that, under appropriate assumptions, it is possible on some probability space to define two normalized sums of iid random variables

$$a(n)^{-1} \Sigma_{i=1}^{n} X_i \quad \text{and} \quad b(n)^{-1} \Sigma_{i=1}^{n} Y_i,$$

with specified distributions for the X's and Y's, so that the difference

$$S_n = b(n)^{-1} \Sigma_{i=1}^{n} Y_i - a(n)^{-1} \Sigma_{i=1}^{n} X_i \tag{1}$$

is virtually non-stochastic. Specifically, it is shown that

$$S_n = h(n) + o_p(1) \quad \text{as} \quad n \to \infty$$

for appropriate constants $h(n)$. (The notation $Z_n = o_p(c_n)$ means $Z_n/c_n \to 0$ in probability.) We shall begin with a discussion of assumptions.

Let X be an unbounded random variable. A positive function $a(\bullet)$ on R^+ (the positive reals) is said to be a <u>norming function</u> for X if

$$\limsup_{x \to \infty} x\Pr(|X| > a(x)) < \infty,$$
$$\liminf_{x \to \infty} x\Pr(|X| > a(x)) > 0, \tag{2}$$

and

$$a(x+t)/a(x) \to 1 \quad \text{as} \quad x \to \infty \quad \text{for each} \quad t > 0. \tag{3}$$

Quite clearly, (2) implies that $a(x) \to \infty$ as $x \to \infty$. In Theorem 1, we assume that $a(\circ)$ is a norming function for X, that $b(\circ)$ is a norming function for Y, and that $a(\circ)$ and $b(\circ)$ are related by the following two conditions:

C1. $a(\circ)/b(\circ)$ is a slowly varying function.

C2. For each $B \in (0,1)$, as $x \to \infty$,

$$x\Pr(\pm Y > B^{-1}b(x)) - o(1) \leq x\Pr(\pm X > a(x)) \leq x\Pr(\pm Y > Bb(x)) + o(1)$$

and

$$x\Pr(\pm X > B^{-1}a(x)) - o(1) \leq x\Pr(\pm Y > b(x)) \leq x\Pr(\pm X > Ba(x)) + o(1).$$

Here, in C2 (and below in A3 and A3'), the interpretation is that the statements in C2 (A3 and A3') are true with all pluses preceding the random variables and with all minuses preceding the random variables. It is further assumed that X and Y satisfy the following three assumptions which are described for X:

A1. $$\limsup_{x\to\infty} \frac{EX^2 I(|X| \leq x)}{x^2 \Pr(|X| > x)} < \infty.$$

A2. $$\sup_{x>1} \frac{\Pr(|X| > Ax)}{\Pr(|X| > x)} \to 0 \quad \text{as} \quad A \to \infty.$$

A3. Given any B and any subsequence of positive integers $n' \to \infty$ for which $n' \Pr(\pm X > Ba(n'))$ is bounded away from zero, there exists for each $B' \in (0,B)$ a further subsequence n'' such that $\Pr(\pm X > B'a(n''))/\Pr(\pm X > Ba(n''))$ is bounded away from one.

The statement of A3 is made somewhat cumbersome in order to make it applicable in a wide variety of situations involving unbalanced tails. For instance, when $n\Pr(X > a(n)) \to 0$, no conclusion is required, and it would be seriously restrictive to require $\Pr(X > B'a(n))/\Pr(X > a(n))$ to be bounded away from one. (The ratio might even assume the form "0/0" for large n.) Although somewhat stronger than A3, assumption A3' below is often true in applications and captures much of the spirit of A3:

A3. Given any $B \in (0,1)$, the ratio $\Pr(\pm X > Bx)/\Pr(\pm X > x)$ is (defined and) bounded away from one for large x.

Note that both A3 and A3 imply that, in certain senses, $\Pr(|X| > y)/\Pr(|X| > x)$ is small for $y > x$.

Assumption A1 simply means that the random variable X is not in the domain of partial attraction of a normal variable. (See Paul Levy (1954), page 113, for a proof.) Integration by parts yields

(4) $$EX^2 I(|X| \leq x) = 2\int_0^x v \Pr(|X| > v)dv - x^2 \Pr(|X| > x).$$

Consequently, A1 is equivalent to

A1'. $$\limsup_{x\to\infty} \frac{\int_0^x v\Pr(|X| > v)dv}{x^2 \Pr(|X| > x)} < \infty.$$

Further, it can be shown that A1 is also equivalent to each of:

A1". For some $\varepsilon > 0$, $x^{-\varepsilon} \int_0^x v \Pr(|X| > v)dv$ is an increasing function of x on R^+.

A1'''. For some $\varepsilon > 0$ and some $M > 1$

$$x^{2-\varepsilon} \Pr(|X| > x) \leq My^{2-\varepsilon} \Pr(|X| > y), \quad y > x > 0.$$

Theorem 1. Suppose X and Y possess norming functions $a(\cdot)$ and $b(\cdot)$, respectively, and that they each satisfy assumptions A1-A3. Further, suppose conditions C1 and C2 are satisfied. Then on some probability space there exist two sequences of iid random variables $X_1, X_2, \ldots$ and $Y_1, Y_2, \ldots$ whose members are distributed as X and Y, respectively, and which satisfy

(5) $$\frac{\Sigma_{i=1}^n Y_i}{b(n)} = \frac{\Sigma_{i=1}^n X_i}{a(n)} + h(n) + o_p(1) \text{ as } n \to \infty$$

for some sequence of constants $h(n)$.

If X is a stable variable of index $\alpha \in (0,2)$ and $a(n) = n^{1/\alpha}$, it immediately follows from (5) that Y is in the domain of attraction of X.

Theorem 1 says something about domains of partial attraction. Let $\mathcal{D}_X$ denote the collection of all (infinitely divisible) random variables Z such that X is in the domain of partial attraction of Z. According to the theory, $\mathcal{D}_X$ is empty, $\mathcal{D}_X$ consists of one type of random variable, or $\mathcal{D}_X$ consists of a non-denumerable set of types (Doeblin (1940), Gnedenko (1940)).

Corollary 1. If X and Y satisfy the hypotheses of Theorem 1 for some choice of norming functions $a(\cdot)$ and $b(\cdot)$, then $\mathcal{D}_X = \mathcal{D}_Y$.

In some applications, it is helpful to be able to replace the condition "$a(\cdot)/b(\cdot)$ is slowly varying" (i.e., condition C1), which appears in Theorem 1, by a condition more directly verifiable from the distribution functions of X and Y. According to the following corollary, this can be done if we assume a little more about the norming function of Y, namely

(6) $$x\Pr(|Y| > b(x)) \text{ is slowly varying in } x,$$

and if we impose the following additional regularity assumption for X:

(7) A positive function $\gamma(\circ)$ on R^+ is slowly varying whenever $x\Pr(|X| > \gamma(x)a(x))$ is slowly varying in X.

Corollary 2. Theorem 1 remains valid if condition C1 is replaced by

(8) $\beta(x) = \Pr(|Y| > x)/\Pr(|X| > x)$ is slowly varying in x,

and the additional assumptions (6) and (7) are valid.

Observe that the conditions (6) and (7), entering into Corollary 2, are asymmetric in X and Y. Quite clearly Corollary 2 would remain valid if (6) and (7) were replaced by

(6') $x \Pr(|X| > a(x))$ is slowly varying in x,

and

(7') A positive function $\gamma(\circ)$ on R^+ is slowly varying whenever $x\ PR(|Y| > \gamma(x)b(x))$ is slowly varying in x.

Theorem 1 is easily specialized to the study of stable (random) variables.

Recall that a random variable X is said to be stable if for each n there are independent random variables $X_1, X_2, \ldots, X_n$ with common distribution that of X, centering constants e(n) and scaling constants $a(n) > 0$ such that

$$(9) \qquad \frac{\Sigma_{i=1}^{n} X_i}{a(n)} - e(n)$$

has the distribution of X. It is well known that, corresponding to each properly scaled non-normal stable variable X, there exists a pair (α, p), $0 < \alpha < 2$, $0 \le p \le 1$, such that

$$(10) \qquad \Pr(|X| > x) \sim 1/x^{\alpha}, \quad \Pr(X > x) \sim p/x^{\alpha} \quad \text{as } x \to \infty.$$

Moreover, for such a random variable, a(n) in (9) must assume the form $n^{1/\alpha}$. Observe, by direct substitution, that $a(x) = x^{1/\alpha}$ is a norming function for any random variable X satisfying (10). Any stable variable X which (as a result of proper scaling) satisfies (10) will be said to be of type (α,p). If the word "type" were to be used as Loève does (1963, page 202), then each pair (α,p) would correspond to exactly one type of stable law. This fact explains our terminology.

Theorem 2. Let X be a stable variable of type (α,p) and Y be an unbounded random variable with a distribution of the form

$$\Pr(|Y| > x) = \beta(x)/x^{\alpha}, \quad x > 0, \tag{11}$$

$$\Pr(Y > x)/\Pr(|Y| > x) \to p \quad \text{as} \quad x \to \infty, \tag{12}$$

where $0 < \alpha < 2$, $0 \leq p \leq 1$, and $\beta(x)$ is a slowly varying function. Then, on some probability space, there exist two sequences of iid random variables $X_1, X_2, \ldots$ and $Y_1, Y_2, \ldots$ whose members are distributed as X and Y respectively, and which satisfy

$$\frac{\Sigma_{i=1}^{n} Y_i}{b(n)} = \frac{\Sigma_{i=1}^{n} X_i}{n^{1/\alpha}} + h(n) + o_p(1) \quad \text{as} \quad n \to \infty \tag{13}$$

for some positive function $b(x)$ and some sequence of centering constants $h(n)$. Moreover, (13) holds for each choice of $b(x)$ satisfying

$$\frac{x\beta(b(x))}{b^{\alpha}(x)} \to 1 \quad \text{as} \quad x \to \infty. \tag{14}$$

Theorem 2 yields the well known classical sufficient conditions for weak convergence of normed sums to a stable law.

Corollary 3. Under the hypotheses of Theorem 2, there exist constants $b(n) > 0$ and $g(n)$ such that

$$\frac{\Sigma_{i=1}^{n} Y_i}{b(n)} - g(n) \xrightarrow{L} X, \tag{15}$$

where $Y_1, Y_2, \ldots$ are iid random variables distributed as Y. Moreover, (15) holds for each positive function $b(x)$ satisfying (14).

Corollary 3 is the strongest possible result in the direction it is stated. That is, a random variable Y is in the domain of attraction of a stable random variable X of type (α,p) $(0<\alpha<2,$ $0 \leq p \leq 1)$ iff Y satisfies the assumptions stated in the corollary.

We have tried unsuccessfully to obtain the converse of Corollary 3 by probabilistic methods. Of course, a proof of the converse based on Fourier transforms is well known.

3. Domain of Normal Attraction of the Cauchy Law.

The following theorem describes a very special case of the weak invariance principle of Theorem 1 but, at the same time, with a minimum of computations, yields an interesting convergence in law result.

Let X be a Cauchy random variable (centered at zero and) scaled so that

$$\Pr(|X| > x) \sim \frac{1}{x} \quad \text{as} \quad x \to \infty ,$$

and Y be a symmetric random variable satisfying

$$\Pr(|Y| > x) \sim \frac{1}{x} \quad \text{as} \quad x \to \infty.$$

Theorem 3. On some probability space, there exist two sequences of iid random variables $X_1, X_2, \ldots$ and $Y_1, Y_2, \ldots$ whose members are distributed as X and Y, respectively, and which satisfy

(16) $$\Sigma_{i=1}^{n} Y_i = \Sigma_{i=1}^{n} X_i + o_p(n) \quad \text{as} \quad n \to \infty.$$

Since $n^{-1}\Sigma_{i=1}^{n} X_i$ is distributed as X, the following corollary is immediate.

Corollary 4. If $Y_1, Y_2, \ldots$ are iid and distributed as Y, then

$$n^{-1}\Sigma_{i=1}^{n} Y_i \xrightarrow{L} X.$$

Proof of Theorem 3. Let $U_1, U_2, \ldots$ be iid uniform variables on $(0,1)$, and let F and G denote the (right continuous) distribution functions of X and Y respectively. Set

$$X_i = F^{-1}(U_i), \quad Y_i = G^{-1}(U_i), \quad i \geq 1.$$

Here F^{-1} is defined formally by

(17) $$F^{-1}(u) = \min\{x : F(x) \geq u\}, \quad u \in (0,1) .$$

Since

(18) $$F^{-1}(u) \leq x \quad \text{iff} \quad u \leq F(x),$$

the distribution function of X_1 is F. Let

$$Z_i = Y_i - X_i \ (i \geq 1) \quad \text{and} \quad Z = Z_1.$$

Equation (16) is identical to $n^{-1}\Sigma_{i=1}^n Z_i \xrightarrow{P} 0$ as $n \to \infty$, which by the classical degenerate convergence theorem, is equivalent (since Z is symmetric) to

(19) $$n \Pr(|Z| > n) \to 0 \quad \text{as} \quad n \to \infty,$$

(20) $$n^{-1} \operatorname{Var} ZI(|Z| \leq n) \to 0 \quad \text{as} \quad n \to \infty .$$

Since $\operatorname{Var} ZI(|Z| \leq n) \leq 2\int_0^n v\Pr(|Z| > v)dv$ (recall (4)), condition (19) implies condition (20).

It is almost immediate that

$$F^{-1}(u) = \frac{1+o(1)}{2(1-u)}, \quad G^{-1}(u) = \frac{1+o(1)}{2(1-u)} \quad \text{as} \quad u \to 1,$$

and

$$F^{-1}(u) = -\frac{1+o(1)}{2u}, \quad G^{-1}(u) = -\frac{1+o(1)}{2u} \quad \text{as} \quad u \to 0.$$

Hence

$$G^{-1}(u) - F^{-1}(u) = o((1-u)^{-1}) \quad \text{as} \quad u \to 1 \quad \text{and} = o(u^{-1}) \quad \text{as } u \to 0.$$

In as much as $Z = G^{-1}(U_1) - F^{-1}(U_1)$, condition (19) follows immediately: For any fixed $\varepsilon > 0$ and sufficiently large n,

$$\Pr(|Z| > n) \leq \Pr(\tfrac{\varepsilon}{1-U_1} > n) + \Pr(\tfrac{\varepsilon}{U_1} > n) = 2\varepsilon/n. \qquad \square$$

It is a simple matter to replace X by any non-normal stable (random) variable in Theorem 3. That is, the classical conditions for a random variable Y to be in the domain of normal attraction of a (non-normal) stable variable can easily be shown to be sufficient using a theorem similar to Theorem 3.

If one were to try to prove the classical central limit theorem using the above approach, one would have to establish

$$\Sigma_{i=1}^n (Z_i - EZ)/\sqrt{n} \xrightarrow{P} 0 \quad \text{as} \quad n \to \infty .$$

But this would require $\operatorname{Var} Z$ to be zero, which is of course too strong. It is from considerations such as this that we were led to work with random variables which are outside the domain of of partial attraction of a normal random variable.

REFERENCES

[1] Csörgö, M. and Révész, P. (1975). A new method to prove Strassen type laws of invariance principle. I. Z. Wahrscheinlichkeitstheorie Verv. Gebiete 31, 255-259.

[2] Doeblin, W. (1940). Sur l' ensemble de pruissances d' une loi de probabilité. Studia Math. 9, 71-96.

[3] Gnedenko, B. V. (1940) (in Russian). Some theorems on the powers of distribution functions. Uchenye Zapiski Moskov. Gos. Univ. Mathematika 45, 61-72.

[4] Komlós, J., Major, P. and Tusnády, G. (1975). An approximation of partial sums of independent RV's and DF. I. Z. Wahrscheinlichkeitstheorie und Verw. Gebiete 32, 111-131.

[5] Lévy, Paul (1954). Théorie de l' Addition des Variables Aléatories, 2nd ed. Paris: Guathier-Villars.

[6] Loève, M. (1963). Probability Theory, 3rd ed. Princeton, New Jersey: Van Nostrand.

[7] Philipp, W., and Stout, W. (1975). Almost sure invariance principles for partial sums of weakly dependent random variables. Amer. Math. Soc. Mem. No. 161. Amer. Math. Soc., Providence, Rhode Island.

[8] Root, D. and Rubin, H. (1973). A probabilistic proof of the normal convergence criterion. Ann. Prob. 1, 867-869.

[9] Strassen, V. (1964). An invariance principle for the law of the iterated logarithm. Z. Wahrscheinlichkeitstheorie und Verw. Gebiete 3, 211-226.

[10] Strassen, V. (1965). Almost sure behavior of sums of independent random variables and martingales. Proc. Fifth Berkeley Symp. Math. Statist. Prob. 2, 315-343.

A NECESSARY CONDITION FOR THE CONVERGENCE OF THE ISOTROPE DISCREPANCY

W. Stute

Ruhr University, Math. Inst. NA

D-4630 Bochum, West Germany

Summary. Given a sequence $(\xi_i)_{i\in\mathbb{N}}$ of independent identically distributed (i.i.d.) $\mathbb{R}^k$-valued random vectors with distribution $\mu = Q_{\xi_1}$, the isotrope discrepancy $D_n^\mu(\omega)$ is defined by $D_n^\mu(\omega):=\sup_{C\in\mathcal{C}_k} |\mu_n^\omega(C)-\mu(C)|$, where μ_n^ω denotes the empirical probability distribution and the supremum is taken over the class $\mathcal{C}_k$ of all convex measurable subsets of $\mathbb{R}^k$. In the present paper it is proved that $\mu_c(e(C))=0$ for all $C\in\mathcal{C}_k$ whenever $D_n^\mu(\omega)\to 0$ a.s., and where $e(C)$ denotes the set of all extreme points of C.

Introduction and results. Let $(\xi_i)_{i\in\mathbb{N}}$ be a sequence of i.i.d. $\mathbb{R}^k$-valued random vectors on some probability space $(\Omega,\mathcal{F},\mathbb{P})$ with probability distribution $\mu=Q_{\xi_1}$. Then, if $\mu_n^\omega := \frac{1}{n}\sum_{i=1}^n \varepsilon_{\xi_i(\omega)}$, $n\in\mathbb{N}$, $\omega\in\Omega$, denotes the empirical distribution pertaining to $\xi_1(\omega),\ldots,\xi_n(\omega)$, the isotrope discrepancy (at stage n) is defined by

$$D_n^\mu(\omega) := \sup_{C\in\mathcal{C}_k} |\mu_n^\omega(C)-\mu(C)|,$$

where the supremum is taken over the class $\mathcal{C}_k$ of all convex measurable subsets of $\mathbb{R}^k$. Obviously, the number $D_n^\mu(\omega)$ can be regarded as describing the imperfection of the distribution (w.r.t. μ) of the first n points of the sequence $(\xi_i(\omega))_{i\in\mathbb{N}}$ over $\mathbb{R}^k$. S.K. Zaremba [5] showed that this type of discrepancy has interesting applications in the theory of numerical integration. He proved that, in the case of $\mu=\lambda_k$ (=Lebesgue measure on the unit cube I^k), for each $C\in\mathcal{C}_k\cap I^k$ and all $f:I^k\to\mathbb{R}$, which are sufficiently smooth, one has

$$\left|\int_C f\,d\mu_n^\omega - \int_C f\,d\mu\right| \le V(f)D_n^\mu(\omega),$$

where V(f) is some constant depending only on f.

Thus, in order to obtain an upper bound for the error of integration as $n\to\infty$, one has to investigate the limit behaviour of $D_n^\mu(\omega)$, $n\to\infty$. R.R. Rao [3] proved that for each μ fulfilling the following condition

$$(+) \quad \sup_{C\in\mathcal{C}_k} \mu_c(\partial C)=0$$

one obtains that $D_n^\mu(\omega)\to 0$ as $n\to\infty$ $\mathbb{P}$-almost surely (where μ_c denotes the nonatomic part of μ and ∂C is the boundary of C).

On the other hand he showed that a condition like (+) cannot be dispensed with in general in the case $k \geq 2$. For example, let μ be the uniform distribution on $S^1 = \{(x_1,x_2)\in\mathbb{R}^2 : x_1^2+x_2^2=1\}$. Then, if $C_n(\omega)$ denotes the convex hull of the sample $(\xi_1(\omega),\ldots,\xi_n(\omega))$, $\omega\in\Omega$, $n\in\mathbb{N}$, then $\mu(C_n(\omega))=0$ and $\mu_n^{\omega}(C_n(\omega))=1$, whence $D_n^{\mu}(\omega)=1$.

The following result is a partial converse to Rao's theorem. For a comprehensive account on similar results we refer to Gaenssler and Stute [1] , [2] . Recall that for $C\in\mathcal{C}_k$ a point $x\in C$ is an extreme point of C if and only if $C\setminus\{x\}$ is convex. Let $e(C)$ denote the set of all extreme points of C.

Theorem. Suppose that with probability one

(1) $D_n^{\mu}(\omega)\to 0$ as $n\to\infty$.

Then (2) $\sup_{C\in\mathcal{C}_k} \mu_c(e(C)) = 0.$

Furthermore (1) and (2) are equivalent in the case $k=1,2$.

Proof of the Theorem.

To prove the necessity of (2) the following lemma will be useful.

Lemma. Suppose that for given $q\in\mathbb{N}$ there exists $C\in\mathcal{C}_k$ and $1\leq i_1<i_2<\ldots<i_q\leq n$ such that

(3) $\xi_{i_j} := \xi_{i_j}(\omega)\in e(C)$

and (4) $\mu(\{\xi_{i_j}\}) = 0 \quad j=1,\ldots,q.$

Then $\quad D_n^{\mu}(\omega) \geq \frac{q}{2n}.$

Proof. Put $B_n(\omega) := \{\xi_{i_j}(\omega) : j=1,\ldots,q\}$ and let $C_n(\omega)$ denote the convex hull of $B_n(\omega)$. Since by (3) $\xi_{i_j}\in e(C)$ it follows that $\xi_{i_j}\in e(C_n(\omega))$, $j=1,\ldots,q$, whence $A_n(\omega):=C_n(\omega)\setminus B_n(\omega)\in\mathcal{C}_k$. Apply (4) to obtain

$$D_n^{\mu}(\omega) \geq |\mu_n^{\omega}(C_n(\omega))-\mu(C_n(\omega))|$$
$$\geq \mu_n^{\omega}(B_n(\omega))-D_n^{\mu}(\omega) \geq \frac{q}{n} - D_n^{\mu}(\omega),$$ whence the assertion. □

We are now in a position to prove the Theorem:

Let M_o denote the set of all μ-atoms and suppose that for some $C\in\mathcal{C}_k$ $\gamma := \mu_c(e(C))>0$. Apply SLLN to obtain $\Omega_o\in\mathfrak{F}$ with $\mathbb{P}(\Omega_o)=1$ and $n_o(\omega)\in\mathbb{N}$ such that for all $\omega\in\Omega_o$ $\mu_n^{\omega}(e(C)-M_o)\geq\gamma/2>0$, $n\geq n_o(\omega)$. Put $q=q(n):=[\frac{\gamma n}{2}]$, $n\in\mathbb{N}$, where $[x]$, $x\in\mathbb{R}$, denotes the greatest integer $\leq x$. Then by the preceding lemma for all $\omega\in\Omega_o$ and $n\geq n_o(\omega)$ $D_n^{\mu}(\omega)\geq q/2n \geq\gamma/4-1/2n$. This contradicts (1) and completes the first part of the proof. To show that (1) is also necessary for (2) in the case k=1,2 we shall apply uniformity techniques as developed in [2] and [4]. In particular, by Theorem 3.1 in [2], in order to prove (1), it will be sufficient to show that under assumption (2) $\mathcal{C}_k$ is a $(\mu, \mathfrak{B}_k)$-uniformity class. To this extent, let μ_d denote the atomic part of μ, i.e. $\mu = \mu_d + \mu_c$. Since both (1) and (2) are obviously true in the case k=1, we may assume w.l.o.g. that k=2. Then by the finiteness of μ_c there are at most countably many straight lines L_i, $i \in T \subset \mathbb{N}$, in $\mathbb{R}^2$ such that $\mu_c(L_i) > 0$. Put $\mu_i := \text{rest}_{L_i} \mu_c$, $i \in T$, and $\mu_o := \text{rest}_{L_o} \mu_c$, where $L_o := \mathbb{R}^2 \setminus (\bigcup_{i \in T} L_i)$. Then clearly $\mu = \mu_d + \mu_o + \sum_{i \in T} \mu_i$. Since ∂C has finite length for each bounded $C \in \mathcal{C}_k$ it follows from standard arguments, that one can find countably many segments S_i, $i \in T' \subset \mathbb{N}$, such that $\partial C \setminus e(C) \subset \bigcup_{i \in T'} S_i$. Thus by the construction of μ_o we obtain from (2) $\mu_o(\partial C) = 0$. For arbitrary $C \in \mathcal{C}_k$ the same follows from $\partial C \subset \bigcup_{n \in \mathbb{N}} \partial(C \cap K_n)$, where K_n denotes the ball with centre $\underline{0} \in \mathbb{R}^2$ and radius $n \in \mathbb{N}$. Hence $\sup_{C \in \mathcal{C}_k} \mu_o(\partial C) = 0$. Now, as was shown in [2], Theorem 2.13, the last condition implies that $\mathcal{C}_k$ is a $(\mu_o, \mathfrak{B}_k)$-uniformity class. Since the same is true for μ_d and μ_i, $i \in T$ (cf. [2], Remark 2.6, Theorem 2.9 and Lemma 2.8), we may apply [2], Lemma 2.5 in order to ensure that $\mathcal{C}_k$ is a $(\mu, \mathfrak{B}_k)$-uniformity class. This completes the proof of the Theorem. □

To show that (1) and (2) are no longer equivalent in the case k=3, let μ be the normalized two-dimensional Lebesgue measure on the envelope E of some cone M. We are going to prove that $\sup_{C \in \mathcal{C}_k} \mu(e(C)) = 0$. Indeed, for given $x \in E$ let g_x denote the straight line joining x and the vertex of M. Then by definition of e(C) card $(e(C) \cap g_x) \leq 2$, so that by Fubini's theorem $\mu(e(C)) = 0$. As to the validity of (1) we remark that the same arguments as in Rao's example apply in order to show that $D_n^{\mu}(\omega) = 1$ $\mathbb{P}$-a.s. for all $n \in \mathbb{N}$.

REFERENCES

[1] Gaenssler, P. and Stute, W. (1975/76). A survey on some results for empirical processes in the i.i.d. case. RUB-Preprint Series No. 15.

[2] Gaenssler, P. and Stute, W. (1976). On uniform convergence of measures with applications to uniform convergence of empirical distributions. In this volume.

[3] Rao, R.R. (1962). Relations between weak and uniform convergence of measures with appl. Ann. Math. Statist. 33, 659-680.

[4] Stute, W. (1976). On a generalization of the Glivenko-Cantelli theorem. Z. Wahrscheinlichkeitstheorie verw. Gebiete 35, 167-175

[5] Zaremba, S.K. (1970). La discrépance isotrope et l'intégration numérique. Ann. Mat. Pura Appl. (IV) 87, 125-135.

TWO EXAMPLES CONCERNING UNIFORM CONVERGENCE OF MEASURES w.r.t. BALLS IN BANACH SPACES

Flemming Topsøe, Richard M. Dudley, Jørgen Hoffmann-Jørgensen

The University of Copenhagen, Massachusetts Institute of Technology, The University of Aarhus.

Summary Two examples show that certain uniform convergence properties related to the Glivenko-Cantelli theorem - properties which are known to hold in Euclidean spaces - need not hold in general Banach spaces.

This research is supported by the Danish Natural Science Research Council.

<u>Introduction</u> $(X, \|\cdot\|)$ denotes a separable Banach space and $\mathcal{B}(X)$ the Borel σ-field on X. Let $\mathcal{A}$ be a subclass of $\mathcal{B}(X)$ and let μ be a probability measure on $\mathcal{B}(X)$. μ is $\mathcal{A}$-<u>continuous</u> if $\mu(\partial A) = 0$ for all $A \in \mathcal{A}$. $\mathcal{A}$ is an <u>ideal uniformity class</u> if

$$\lim_{n\to\infty} \sup_{A\in\mathcal{A}} |\mu_n(A)-\mu(A)| = 0$$

for every $\mathcal{A}$-continuous μ and any sequence $(\mu_n)_{n\geq 1}$ of probability measures converging weakly to μ. Further, we say that the <u>Glivenko-Cantelli assertion holds for $\mathcal{A}$ with theoretical distribution</u> μ if

$$\lim_{n\to\infty} \sup_{A\in\mathcal{A}} |\mu_n^{\omega}(A)-\mu(A)| = 0 \quad \text{a.e.} \quad [\omega] ,$$

where μ_n^{ω} denotes the empirical measures pertaining to the theoretical distribution μ. If $\mathcal{A}$ is an ideal uniformity class, it follows by a result of Varadarajan [5], that the Glivenko-Cantelli assertion holds for $\mathcal{A}$ and any $\mathcal{A}$-continuous probability measure μ.

We shall investigate the class $\mathcal{B}_1$ of all closed balls

$$B[x,r] = \{y: \|x-y\| \leq r\}$$

of radius $r \leq 1$. It is known that $\mathcal{B}_1$ is an ideal uniformity class if X is finite dimensional or if $X = l^p$ with $1 \leq p < \infty$, cf. [3]. Recently, it was proved by Elker [1] that in case X is a finite dimensional Euclidean space, the Glivenko-Cantelli assertion holds for $\mathcal{B}_1$ and <u>any</u> probability measure μ (it is even unnecessary to put a bound on the radia of the balls considered). The work of Elker is unpublished; a proof can be deduced from [2] - it should also be possible to base a proof on [4], however, it is not so obvious how to solve the combinatorial problem you are led to consider using this source.

The two examples which follow are negative results and should be seen in the light of the above mentioned positive results.

<u>Balls in L^1</u>. Denote by L^1 the L^1-space over $[0,1]$ with Lebesgue measure. Define, for $n \geq 1$, $A_n^+ \subseteq [0,1]$ and $A_n^- \subseteq [0,1]$ by

$$A_n^+ = \bigcup_{\nu=0}^{2^{n-1}-1} [2\nu\cdot 2^{-n}, (2\nu+1)\cdot 2^{-n}],$$

$$A_n^- = \bigcup_{\nu=0}^{2^{n-1}-1} [(2\nu+1)\cdot 2^{-n}, (2\nu+2)\cdot 2^{-n}]$$

(i.e. $1_{A_n^+} - 1_{A_n^-}$ is the n'th Rademacher function).

It is elementary to check, that for any $f \in L^1$, we have

(1) $$2\int_{A_n^+} f(t)dt \to \int_{[0,1]} f(t)dt,$$

(2) $$2\int_{A_n^-} f(t)dt \to \int_{[0,1]} f(t)dt .$$

<u>Lemma</u>. <u>Let</u> $f \in L^1$ <u>and define</u> $f_n; n \geq 1$ <u>by</u>

$$f_n = f \cdot 1_{A_n^+} - f \cdot 1_{A_n^-} .$$

<u>Then, for any</u> $g \in L^1$, <u>we have</u>

$$\| f_n - g \| \to \| \max(|f|,|g|) \| .$$

<u>Proof</u>. By (1) and (2) we find:

$$\begin{aligned} \| f_n - g \| &= \int_{A_n^+} |f-g| + \int_{A_n^-} |f+g| \\ &\to \tfrac{1}{2}\int_{[0,1]} |f-g| + \tfrac{1}{2}\int_{[0,1]} |f+g| \\ &= \| \max(|f|,|g|) \| . \quad \text{I} \end{aligned}$$

Recall that for a closed subset F of a Banach space X and for a sequence $(F_n)_{n\geq 1}$ of closed subsets of X, we call F the <u>topological limit</u> of $(F_n)_{n\geq 1}$, and write $F_n \to F$, if F coincides with the set of points whose distance to F_n converges to 0, and if F also coincides with the set of points whose distance to a subsequence of (F_n) converges to 0. For the role of this notion in connection with the uniformity properties under discussion, we refer to [3] .

<u>Proposition 1</u>. <u>Let</u> $f \in L^1$ <u>and assume that</u> $\| f \| \leq 1$. <u>Then there exists a sequence</u> $(B_n)_{n\geq 1}$ <u>of balls in</u> $\mathcal{B}_1$ <u>such that</u>

$$\partial B_n \to \{g \in L^1 : |g| \leq |f|\} .$$

<u>Proof</u>. Define f_n as in the lemma. The lemma shows that

$$\partial B[f_n, \| f \|] \to \{g \in L^1 : \| \max(|f|,|g|) \| = \| f \|\} .$$

The result now follows from the obvious equivalence

$$\|\max(|f|,|g|)\| = \|f\| \Leftrightarrow |g| \leq |f| \ . \quad \blacksquare$$

<u>Theorem 1</u>. *In* L^1, $\mathcal{B}_1$ *is not an ideal uniformity class.*

<u>Proof</u>. It is easy to see that there exists a $\mathcal{B}_1$-continuous measure μ for which

$$\mu(\{g \in L^1 : |g| \leq 1\}) > 0$$

- for instance, you may consider Wiener measure. This fact in connection with Proposition 1 and Lemma 2, (iv) of [3] shows that $\mathcal{B}_1$ is not an ideal uniformity class. $\blacksquare$

<u>Remark</u>. In the terminology of [3], we believe that in L^1, $\mathcal{B}_1$ is not a μ-uniformity class for any non-zero μ. In [3] this kind of pathological behaviour was shown to hold in the space c_0.

<u>Balls in Hilbert space.</u>

<u>Proposition 2</u>. *Let* X *be a separable Hilbert space. There exists a subset* K *of* X *homeomorphic to the unit interval* $[0,1]$ *and such that, for any finite subset* K_0 *of* K, *there exists* $y \in X$ *and* $0 < r \leq 1$ *such that*

$$\|y-x\| = r \quad \textit{for all} \quad x \in K_0$$

and such that

$$\|y-x\| > r \quad \textit{for all} \quad x \in K \setminus K_0 .$$

<u>Proof</u>. Consider Brownian motion as a curve $t \to x(t)$ in Hilbert space. Concretely, we may take as X the L^2-space over $[0,1]$ with Lebesgue measure, and define

$$x(t) = 1_{[0,t]} \ ; \quad t \in [0,1].$$

Indeed, we find that

$$\langle x(t),x(s)\rangle = \min(t,s).$$

We put

$$K = \{x(t): t \in [0,1]\}.$$

Clearly, K is homeomorphic to $[0,1]$.

For the calculations to follow we need the facts that for $s,t \in [0,1]$

$$\| x(t)-x(s) \|^2 = |t-s|$$

and that the increments are orthogonal.

Let $t_{2\nu-1}$; $\nu = 1,2,\cdots,n$ be n given numbers with

$$0 \leq t_1 < t_3 < \cdots < t_{2n-1} \leq 1.$$

Put

$$t_{2\nu} = \tfrac{1}{2}(t_{2\nu-1}+t_{2\nu+1}); \; \nu = 1,2,\cdots,n-1.$$

Consider the point y and the number r defined by

$$y = \Sigma_{\nu=1}^{n} x(t_{2\nu-1}) - \Sigma_{\nu=1}^{n-1} x(t_{2\nu})$$

$$r = \{\tfrac{1}{2}(t_{2n-1}-t_1)\}^{\frac{1}{2}} .$$

For every $t \in [0,1]$, we shall calculate $\|y-x(t)\|$. Consider the case that, for some k with $1 \leq k \leq n-1$ we have

$$t_{2k-1} \leq t \leq t_{2k} .$$

We find:

$$\begin{aligned}\| y-x(t) \|^2 &= \| \Sigma_{\nu=1}^{k-1}(x(t_{2\nu-1})-x(t_{2\nu})) + (x(t_{2k-1})-x(t)) + \\ &\qquad + \Sigma_{\nu=k}^{n-1}(x(t_{2\nu+1})-x(t_{2\nu})) \|^2 \\ &= \Sigma_{\nu=1}^{k-1}(t_{2\nu}-t_{2\nu-1}) + (t-t_{2k-1}) + \Sigma_{\nu=k}^{n-1}(t_{2\nu+1}-t_{2\nu}) \\ &= \tfrac{1}{2}\Sigma_{\nu=1}^{k-1}(t_{2\nu+1}-t_{2\nu-1}) + (t-t_{2k-1}) + \Sigma_{\nu=k}^{n-1}(t_{2\nu+1}-t_{2\nu-1}) \\ &= r^2 + (t-t_{2k-1}) .\end{aligned}$$

This shows that $\|y-x(t)\| > r$ unless $t = t_{2k-1}$ in which case $\|y-x(t)\| = r$.

The same result is obtained when $t_{2k} \leq t \leq t_{2k+1}$ for some k with $1 \leq k \leq n-1$, and also when $0 \leq t \leq t_1$ or when $t_{2n-1} \leq t \leq 1$.

We have thus seen that if $t \in [0,1]$ is one of the points $t_{2\nu-1}$; $\nu = 1,\cdots,n$ then $\|y-x(t)\| = r$ and if t is any other point in $[0,1]$, then $\|y-x(t)\| > r$. As $r \leq 1$, K has all the desired properties. ∎

Theorem 2. *If X is a separable Hilbert space, there exists a probability measure μ on $\mathcal{B}(X)$ such that the Glivenko-Cantelli assertion fails for $\mathcal{B}_1$ with theoretical distribution μ.*

Proof. Let μ be a non-atomic probability measure concentrated on the set K of Proposition 2. Let (μ_n^ω) be the empirical proces pertaining to μ. Consider some ω and some n. Denote by F the support of μ_n^ω. Then F is a finite subset of K. By Proposition 2, there exists $B \in \mathcal{B}_1$ such that $B \cap K = F$. Thus $\mu(B) = 0$ and $\mu_n^\omega(B) = 1$, hence the Glivenko-Cantelli assertion fails. ∎

Remark. By results already referred to, μ can not be $\mathcal{B}_1$-continuous. It is not surprising then, that the measure μ constructed in the proof has support contained in a sphere. Indeed, with notation as in Proposition 2, we find that $\|x(t)-\frac{1}{2}\| = \frac{1}{2}$ for all $t \in [0,1]$.

REFERENCES

[1] Elker,J.: Unpublished "Diplomarbeit" from the Ruhr university, Bochum 1975.

[2] Topsøe, F.: On the Glivenko-Cantelli Theorem. Z. Wahrscheinlichkeitsrechnung verw. Geb. 14, 239-250 (1970).

[3] Topsøe, F.: Uniformity in weak convergence w.r.t. balls in Banach spaces. Math. Scand. 38, 148-158 (1976).

[4] Vapnik, V.N., and Červonenkis, A.Ya.: On the uniform convergence of relative frequencies of events to their probabilities. Theor. Probability Appl. 16, 264-280 (1971).

[5] Varadarajan, V.S.: On the convergence of probability distributions, Sankyã 19, 23-26 (1958).

Vol. 399: Functional Analysis and its Applications. Proceedings 1973. Edited by H. G. Garnir, K. R. Unni and J. H. Williamson. II, 584 pages. 1974.

Vol. 400: A Crash Course on Kleinian Groups. Proceedings 1974. Edited by L. Bers and I. Kra. VII, 130 pages. 1974.

Vol. 401: M. F. Atiyah, Elliptic Operators and Compact Groups. V, 93 pages. 1974.

Vol. 402: M. Waldschmidt, Nombres Transcendants. VIII, 277 pages. 1974.

Vol. 403: Combinatorial Mathematics. Proceedings 1972. Edited by D. A. Holton. VIII, 148 pages. 1974.

Vol. 404: Théorie du Potentiel et Analyse Harmonique. Edité par J. Faraut. V, 245 pages. 1974.

Vol. 405: K. J. Devlin and H. Johnsbråten, The Souslin Problem. VIII, 132 pages. 1974.

Vol. 406: Graphs and Combinatorics. Proceedings 1973. Edited by R. A. Bari and F. Harary. VIII, 355 pages. 1974.

Vol. 407: P. Berthelot, Cohomologie Cristalline des Schémas de Caracteristique $p > o$. II, 604 pages. 1974.

Vol. 408: J. Wermer, Potential Theory. VIII, 146 pages. 1974.

Vol. 409: Fonctions de Plusieurs Variables Complexes, Séminaire François Norguet 1970–1973. XIII, 612 pages. 1974.

Vol. 410: Séminaire Pierre Lelong (Analyse) Année 1972–1973. VI, 181 pages. 1974.

Vol. 411: Hypergraph Seminar. Ohio State University, 1972. Edited by C. Berge and D. Ray-Chaudhuri. IX, 287 pages. 1974.

Vol. 412: Classification of Algebraic Varieties and Compact Complex Manifolds. Proceedings 1974. Edited by H. Popp. V, 333 pages. 1974.

Vol. 413: M. Bruneau, Variation Totale d'une Fonction. XIV, 332 pages. 1974.

Vol. 414: T. Kambayashi, M. Miyanishi and M. Takeuchi, Unipotent Algebraic Groups. VI, 165 pages. 1974.

Vol. 415: Ordinary and Partial Differential Equations. Proceedings 1974. XVII, 447 pages. 1974.

Vol. 416: M. E. Taylor, Pseudo Differential Operators. IV, 155 pages. 1974.

Vol. 417: H. H. Keller, Differential Calculus in Locally Convex Spaces. XVI, 131 pages. 1974.

Vol. 418: Localization in Group Theory and Homotopy Theory and Related Topics. Battelle Seattle 1974 Seminar. Edited by P. J. Hilton. VI, 172 pages 1974.

Vol. 419: Topics in Analysis. Proceedings 1970. Edited by O. E. Lehto, I. S. Louhivaara, and R. H. Nevanlinna. XIII, 392 pages. 1974.

Vol. 420: Category Seminar. Proceedings 1972/73. Edited by G. M. Kelly. VI, 375 pages. 1974.

Vol. 421: V. Poénaru, Groupes Discrets. VI, 216 pages. 1974.

Vol. 422: J.-M. Lemaire, Algèbres Connexes et Homologie des Espaces de Lacets. XIV, 133 pages. 1974.

Vol. 423: S. S. Abhyankar and A. M. Sathaye, Geometric Theory of Algebraic Space Curves. XIV, 302 pages. 1974.

Vol. 424: L. Weiss and J. Wolfowitz, Maximum Probability Estimators and Related Topics. V, 106 pages. 1974.

Vol. 425: P. R. Chernoff and J. E. Marsden, Properties of Infinite Dimensional Hamiltonian Systems. IV, 160 pages. 1974.

Vol. 426: M. L. Silverstein, Symmetric Markov Processes. X, 287 pages. 1974.

Vol. 427: H. Omori, Infinite Dimensional Lie Transformation Groups. XII, 149 pages. 1974.

Vol. 428: Algebraic and Geometrical Methods in Topology, Proceedings 1973. Edited by L. F. McAuley. XI, 280 pages. 1974.

Vol. 429: L. Cohn, Analytic Theory of the Harish-Chandra C-Function. III, 154 pages. 1974.

Vol. 430: Constructive and Computational Methods for Differential and Integral Equations. Proceedings 1974. Edited by D. L. Colton and R. P. Gilbert. VII, 476 pages. 1974.

Vol. 431: Séminaire Bourbaki – vol. 1973/74. Exposés 436–452. IV, 347 pages. 1975.

Vol. 432: R. P. Pflug, Holomorphiegebiete, pseudokonvexe Gebiete und das Levi-Problem. VI, 210 Seiten. 1975.

Vol. 433: W. G. Faris, Self-Adjoint Operators. VII, 115 pages. 1975.

Vol. 434: P. Brenner, V. Thomée, and L. B. Wahlbin, Besov Spaces and Applications to Difference Methods for Initial Value Problems. II, 154 pages. 1975.

Vol. 435: C. F. Dunkl and D. E. Ramirez, Representations of Commutative Semitopological Semigroups. VI, 181 pages. 1975.

Vol. 436: L. Auslander and R. Tolimieri, Abelian Harmonic Analysis, Theta Functions and Function Algebras on a Nilmanifold. V, 99 pages. 1975.

Vol. 437: D. W. Masser, Elliptic Functions and Transcendence. XIV, 143 pages. 1975.

Vol. 438: Geometric Topology. Proceedings 1974. Edited by L. C. Glaser and T. B. Rushing. X, 459 pages. 1975.

Vol. 439: K. Ueno, Classification Theory of Algebraic Varieties and Compact Complex Spaces. XIX, 278 pages. 1975

Vol. 440: R. K. Getoor, Markov Processes: Ray Processes and Right Processes. V, 118 pages. 1975.

Vol. 441: N. Jacobson, PI-Algebras. An Introduction. V, 115 pages. 1975.

Vol. 442: C. H. Wilcox, Scattering Theory for the d'Alembert Equation in Exterior Domains. III, 184 pages. 1975.

Vol. 443: M. Lazard, Commutative Formal Groups. II, 236 pages. 1975.

Vol. 444: F. van Oystaeyen, Prime Spectra in Non-Commutative Algebra. V, 128 pages. 1975.

Vol. 445: Model Theory and Topoi. Edited by F. W. Lawvere, C. Maurer, and G. C. Wraith. III, 354 pages. 1975.

Vol. 446: Partial Differential Equations and Related Topics. Proceedings 1974. Edited by J. A. Goldstein. IV, 389 pages. 1975.

Vol. 447: S. Toledo, Tableau Systems for First Order Number Theory and Certain Higher Order Theories. III, 339 pages. 1975.

Vol. 448: Spectral Theory and Differential Equations. Proceedings 1974. Edited by W. N. Everitt. XII, 321 pages. 1975.

Vol. 449: Hyperfunctions and Theoretical Physics. Proceedings 1973. Edited by F. Pham. IV, 218 pages. 1975.

Vol. 450: Algebra and Logic. Proceedings 1974. Edited by J. N. Crossley. VIII, 307 pages. 1975.

Vol. 451: Probabilistic Methods in Differential Equations. Proceedings 1974. Edited by M. A. Pinsky. VII, 162 pages. 1975.

Vol. 452: Combinatorial Mathematics III. Proceedings 1974. Edited by Anne Penfold Street and W. D. Wallis. IX, 233 pages. 1975.

Vol. 453: Logic Colloquium. Symposium on Logic Held at Boston, 1972–73. Edited by R. Parikh. IV, 251 pages. 1975.

Vol. 454: J. Hirschfeld and W. H. Wheeler, Forcing, Arithmetic, Division Rings. VII, 266 pages. 1975.

Vol. 455: H. Kraft, Kommutative algebraische Gruppen und Ringe. III, 163 Seiten. 1975.

Vol. 456: R. M. Fossum, P. A. Griffith, and I. Reiten, Trivial Extensions of Abelian Categories. Homological Algebra of Trivial Extensions of Abelian Categories with Applications to Ring Theory. XI, 122 pages. 1975.

Vol. 457: Fractional Calculus and Its Applications. Proceedings 1974. Edited by B. Ross. VI, 381 pages. 1975.

Vol. 458: P. Walters, Ergodic Theory – Introductory Lectures. VI, 198 pages. 1975.

Vol. 459: Fourier Integral Operators and Partial Differential Equations. Proceedings 1974. Edited by J. Chazarain. VI, 372 pages. 1975.

Vol. 460: O. Loos, Jordan Pairs. XVI, 218 pages. 1975.

Vol. 461: Computational Mechanics. Proceedings 1974. Edited by J. T. Oden. VII, 328 pages. 1975.

Vol. 462: P. Gérardin, Construction de Séries Discrètes p-adiques. »Sur les séries discrètes non ramifiées des groupes réductifs déployés p-adiques«. III, 180 pages. 1975.

Vol. 463: H.-H. Kuo, Gaussian Measures in Banach Spaces. VI, 224 pages. 1975.

Vol. 464: C. Rockland, Hypoellipticity and Eigenvalue Asymptotics. III, 171 pages. 1975.

Vol. 465: Séminaire de Probabilités IX. Proceedings 1973/74. Edité par P. A. Meyer. IV, 589 pages. 1975.

Vol. 466: Non-Commutative Harmonic Analysis. Proceedings 1974. Edited by J. Carmona, J. Dixmier and M. Vergne. VI, 231 pages. 1975.

Vol. 467: M. R. Essén, The Cos $\pi\lambda$ Theorem. With a paper by Christer Borell. VII, 112 pages. 1975.

Vol. 468: Dynamical Systems – Warwick 1974. Proceedings 1973/74. Edited by A. Manning. X, 405 pages. 1975.

Vol. 469: E. Binz, Continuous Convergence on C(X). IX, 140 pages. 1975.

Vol. 470: R. Bowen, Equilibrium States and the Ergodic Theory of Anosov Diffeomorphisms. III, 108 pages. 1975.

Vol. 471: R. S. Hamilton, Harmonic Maps of Manifolds with Boundary. III, 168 pages. 1975.

Vol. 472: Probability-Winter School. Proceedings 1975. Edited by Z. Ciesielski, K. Urbanik, and W. A. Woyczyński. VI, 283 pages. 1975.

Vol. 473: D. Burghelea, R. Lashof, and M. Rothenberg, Groups of Automorphisms of Manifolds. (with an appendix by E. Pedersen) VII, 156 pages. 1975.

Vol. 474: Séminaire Pierre Lelong (Analyse) Année 1973/74. Edité par P. Lelong. VI, 182 pages. 1975.

Vol. 475: Répartition Modulo 1. Actes du Colloque de Marseille-Luminy, 4 au 7 Juin 1974. Edité par G. Rauzy. V, 258 pages. 1975. 1975.

Vol. 476: Modular Functions of One Variable IV. Proceedings 1972. Edited by B. J. Birch and W. Kuyk. V, 151 pages. 1975.

Vol. 477: Optimization and Optimal Control. Proceedings 1974. Edited by R. Bulirsch, W. Oettli, and J. Stoer. VII, 294 pages. 1975.

Vol. 478: G. Schober, Univalent Functions – Selected Topics. V, 200 pages. 1975.

Vol. 479: S. D. Fisher and J. W. Jerome, Minimum Norm Extremals in Function Spaces. With Applications to Classical and Modern Analysis. VIII, 209 pages. 1975.

Vol. 480: X. M. Fernique, J. P. Conze et J. Gani, Ecole d'Eté de Probabilités de Saint-Flour IV–1974. Edité par P.-L. Hennequin. XI, 293 pages. 1975.

Vol. 481: M. de Guzmán, Differentiation of Integrals in R^n. XII, 226 pages. 1975.

Vol. 482: Fonctions de Plusieurs Variables Complexes II. Séminaire François Norguet 1974–1975. IX, 367 pages. 1975.

Vol. 483: R. D. M. Accola, Riemann Surfaces, Theta Functions, and Abelian Automorphisms Groups. III, 105 pages. 1975.

Vol. 484: Differential Topology and Geometry. Proceedings 1974. Edited by G. P. Joubert, R. P. Moussu, and R. H. Roussarie. IX, 287 pages. 1975.

Vol. 485: J. Diestel, Geometry of Banach Spaces – Selected Topics. XI, 282 pages. 1975.

Vol. 486: S. Stratila and D. Voiculescu, Representations of AF-Algebras and of the Group U (∞). IX, 169 pages. 1975.

Vol. 487: H. M. Reimann und T. Rychener, Funktionen beschränkter mittlerer Oszillation. VI, 141 Seiten. 1975.

Vol. 488: Representations of Algebras, Ottawa 1974. Proceedings 1974. Edited by V. Dlab and P. Gabriel. XII, 378 pages. 1975.

Vol. 489: J. Bair and R. Fourneau, Etude Géométrique des Espaces Vectoriels. Une Introduction. VII, 185 pages. 1975.

Vol. 490: The Geometry of Metric and Linear Spaces. Proceedings 1974. Edited by L. M. Kelly. X, 244 pages. 1975.

Vol. 491: K. A. Broughan, Invariants for Real-Generated Uniform Topological and Algebraic Categories. X, 197 pages. 1975.

Vol. 492: Infinitary Logic: In Memoriam Carol Karp. Edited by D. W. Kueker. VI, 206 pages. 1975.

Vol. 493: F. W. Kamber and P. Tondeur, Foliated Bundles and Characteristic Classes. XIII, 208 pages. 1975.

Vol. 494: A Cornea and G. Licea. Order and Potential Resolvent Families of Kernels. IV, 154 pages. 1975.

Vol. 495: A. Kerber, Representations of Permutation Groups II. V, 175 pages. 1975.

Vol. 496: L. H. Hodgkin and V. P. Snaith, Topics in K-Theory. Two Independent Contributions. III, 294 pages. 1975.

Vol. 497: Analyse Harmonique sur les Groupes de Lie. Proceedings 1973–75. Edité par P. Eymard et al. VI, 710 pages. 1975.

Vol. 498: Model Theory and Algebra. A Memorial Tribute to Abraham Robinson. Edited by D. H. Saracino and V. B. Weispfenning. X, 463 pages. 1975.

Vol. 499: Logic Conference, Kiel 1974. Proceedings. Edited by G. H. Müller, A. Oberschelp, and K. Potthoff. V, 651 pages 1975.

Vol. 500: Proof Theory Symposion, Kiel 1974. Proceedings. Edited by J. Diller and G. H. Müller. VIII, 383 pages. 1975.

Vol. 501: Spline Functions, Karlsruhe 1975. Proceedings. Edited by K. Böhmer, G. Meinardus, and W. Schempp. VI, 421 pages. 1976.

Vol. 502: János Galambos, Representations of Real Numbers by Infinite Series. VI, 146 pages. 1976.

Vol. 503: Applications of Methods of Functional Analysis to Problems in Mechanics. Proceedings 1975. Edited by P. Germain and B. Nayroles. XIX, 531 pages. 1976.

Vol. 504: S. Lang and H. F. Trotter, Frobenius Distributions in GL_2-Extensions. III, 274 pages. 1976.

Vol. 505: Advances in Complex Function Theory. Proceedings 1973/74. Edited by W. E. Kirwan and L. Zalcman. VIII, 203 pages. 1976.

Vol. 506: Numerical Analysis, Dundee 1975. Proceedings. Edited by G. A. Watson. X, 201 pages. 1976.

Vol. 507: M. C. Reed, Abstract Non-Linear Wave Equations. VI, 128 pages. 1976.

Vol. 508: E. Seneta, Regularly Varying Functions. V, 112 pages. 1976.

Vol. 509: D. E. Blair, Contact Manifolds in Riemannian Geometry. VI, 146 pages. 1976.

Vol. 510: V. Poènaru, Singularités C^∞ en Présence de Symétrie. V, 174 pages. 1976.

Vol. 511: Séminaire de Probabilités X. Proceedings 1974/75. Edité par P. A. Meyer. VI, 593 pages. 1976.

Vol. 512: Spaces of Analytic Functions, Kristiansand, Norway 1975. Proceedings. Edited by O. B. Bekken, B. K. Øksendal, and A. Stray. VIII, 204 pages. 1976.

Vol. 513: R. B. Warfield, Jr. Nilpotent Groups. VIII, 115 pages. 1976.

Vol. 514: Séminaire Bourbaki vol. 1974/75. Exposés 453 – 470. IV, 276 pages. 1976.

Vol. 515: Bäcklund Transformations. Nashville, Tennessee 1974. Proceedings. Edited by R. M. Miura. VIII, 295 pages. 1976.